Astrophysics and Space Science Proceedings

Volume 56

More information about this series at http://www.springer.com/series/7395

Rouven Essig · Jonathan Feng · Kathryn Zurek
Editors

Illuminating Dark Matter

Proceedings of a Simons Symposium

 Springer

Editors
Rouven Essig
C.N. Yang Institute for Theoretical Physics
Stony Brook University
Stony Brook, NY, USA

Jonathan Feng
Department of Physics and Astronomy
University of California, Irvine
Irvine, CA, USA

Kathryn Zurek
Theoretical Physics Group
Lawrence Berkeley National Laboratory
Berkeley, CA, USA

ISSN 1570-6591 ISSN 1570-6605 (electronic)
Astrophysics and Space Science Proceedings
ISBN 978-3-030-31595-5 ISBN 978-3-030-31593-1 (eBook)
https://doi.org/10.1007/978-3-030-31593-1

This Springer imprint is published by the registered company Springer Nature Switzerland AG
The registered company address is: Gewerbestrasse 11, 6330 Cham, Switzerland

Preface

The identity of dark matter is one of the great scientific mysteries of our time. The field is currently undergoing a transformation. The odds-on favorites from earlier decades, WIMPs—while not excluded—are being increasingly squeezed by the lack of positive signals in direct detection or at the LHC. The other classic candidates, sterile neutrinos and axions, are being reexamined, with qualitatively new possibilities (especially as concerns detection) emerging. In the last decade, the community has dramatically broadened the range of dark matter theories it has studied, motivating new searches, including experiments that are remarkably small, cheap, and fast, but nevertheless provide sensitive probes of these new ideas. At the same time, the era of precision astrophysics and cosmology is placing powerful new constraints on dark matter candidates through the observation of dark matter clustering on small scales, an example of the microscopic dynamics of particles imprinting itself on macroscopic scales. These developments together make clear that the field is unlikely to look very similar in one decade to how it looks today.

The Simons Symposium on Illuminating Dark Matter sought to move the discussion forward at this pivotal point in time. It brought together 23 researchers to take a fresh look at dark matter. The meeting's participants spanned the wide range of fields that are now connected with dark matter, with interests ranging from astrophysics and cosmology to particle physics, and even to condensed matter and atomic physics. Each participant contributed a 30-minute talk, which was followed by ample time for discussion. Participants were encouraged to consider different time periods (short, medium, long) and how their subfields could progress during each of these periods. In the afternoon, the Symposium included organized, but free-ranging, discussions, in which participants discussed hot and controversial topics in more depth and with more explanation than is usual at conferences. In this way, the Symposium, which was simultaneously informal and intense, bridged many divides, for example, between astrophysicists and particle physicists, and between theorists and experimentalists.

A major topic of discussion at the Symposium was the extent to which the small-scale structure of the distribution of dark matter in the universe could constrain or motivate new particle dark matter properties. Manoj Kaplinghat argued that a variety of disagreements between small-scale structure simulations and observations, particularly the diversity of halo distributions, could be taken as evidence for strongly self-interacting dark matter. Alyson Brooks, Phil Hopkins, and Julio Navarro showed the promise—and limitations—of simulations for constraining dark matter interactions and structure. Through debate and discussion, it became clear that baryon and dark matter dynamics are still difficult to disentangle in their impact on the dark matter distribution of halos like the Milky Way and dwarf satellites. The field of simulations is not yet at the point that robust upper limits on dark matter self-interactions (in the presence of baryons) can be quoted, but it was emphasized in the discussion that a sign of precision in simulations would be robust upper bounds on dark matter interaction strengths that monotonically decrease with time. Jo Bovy showed that a particularly promising avenue for determining the clustering of dark matter on the smallest scales are narrow stellar streams in the halo of the Milky Way. Neal Dalal and Neal Weiner discussed how sensitive new gravitational lensing measurements could also dramatically increase our understanding of the small-scale structure of dark matter. Because different dark matter candidates leave different imprints in the small-scale structure, one might be able to differentiate between dark matter candidates this way.

Particularly interesting are light dark matter particles and dark sectors with masses in the meV to GeV range, which could explain many puzzles. Moreover, several mechanisms exist that can naturally generate dark matter with the correct relic abundance in this mass range. Recent years have seen an explosion of new ideas to detect such light dark matter particles. These ideas have often emerged from theoretical physicists thinking across several disciplines, including particle physics, condensed matter physics, cosmology, and atomic, molecular, and optical physics. But these ideas also require expert experimentalists and instrumentalists to sharpen these ideas and bring them to fruition. These developments are allowing physicists to explore vast new regions of dark matter parameter space.

Several concrete experimental proposals have been developed to tackle the detection of these theories of dark matter. Rouven Essig discussed techniques, focused on electronic excitation and ionization in atoms and in semiconductors, to detect dark matter with mass in the MeV to GeV range. Javier Tiffenberg showed that an experiment using CCDs based on these ideas was near to being realized. Rafael Lang emphasized that, although it is important to pursue hidden sector dark matter, the WIMP endures and it is important to pursue the detection of this candidate all the way down to the neutrino background. Adam Ritz discussed how sub-MeV dark matter can be accelerated in the Sun to higher velocities and then be probed in direct-detection experiments on Earth. Roni Harnik discussed how dark matter detectors can be used to probe nonstandard neutrino interactions, and how neutrino detectors can probe novel dark matter candidates.

In addition to detecting the presence of dark matter through its scattering off normal matter, dark matter can also be discovered by producing it in particle accelerators and colliders. This approach has been known for a long time, but the growth of interest in light dark matter has opened a vast array of new possibilities. Bertrand Echenard presented a comprehensive review of sub-GeV dark matter searches at accelerators, surveying the many new initiatives around the world with an emphasis on the proposed LDMX experiment, searching for invisible dark mediator decays with unprecedented sensitivity. Mauro Raggi focused on the possibility of discovering dark matter and dark sectors with positron beams, for example, through resonant searches for $e^+ e^- \to X \to e^+ e^-$, where X is a new particle. In particular, PADME is starting to take data and may be able to definitively test many new physics explanations of the $6.8\,\sigma$ beryllium 8 anomaly. Jonathan Feng reviewed motivations for the new emphasis on light dark sectors, emphasized the genericity of non-renormalizable portal interactions, and described FASER, a small and inexpensive proposed experiment that will extend the LHC's sensitivity to light and weakly interacting new particles.

While much of the discussion focused on the myriad of new experiments and probes being proposed (by both theorists and experimentalists), several talks emphasized that rich model building avenues remain. Kathryn Zurek considered a simple dark sector of asymmetric dark matter in the absence of a dark analogue of electromagnetism, and showed that huge bound states, as heavy as 10^{19} GeV, are observationally and cosmologically viable and give rise to unique experimental signatures. Tomer Volansky discussed what the properties needed for dark matter to explain the recent 21 cm observation and also how dissipative dark matter could enhance the growth rate of supermassive black holes. Josh Ruderman sought to build a simple hidden sector dark photon model that also could explain the recent 21 cm observation. Jessie Shelton focused on cosmological and terrestrial signatures of hidden sectors with a dark radiation bath.

Finally, new progress was reported on a number of classic dark matter candidates. The recent observations of gravitational waves by LIGO have renewed interest in primordial black hole (PBH) dark matter. Bernard Carr reviewed the fascinating history of PBHs, the three open windows at intermediate sublunar, and asteroid masses, and stressed the interesting implications of PBHs even if they account for only some of the dark matter. Alex Kusenko presented a new paradigm for production of black holes in the early universe and noted that PBH dark matter can contribute to r-process nucleosynthesis, as well as lead to striking signatures, such as kilonova without gravitational wave counterparts and fast radio bursts. Aaron Chou reviewed the classic motivations for axion dark matter, presented "hot off the press" results from ADMX that have reached the DFSZ limit for masses around 2 μeV, and discussed a number of new ideas from quantum metrology to enable higher mass searches. Finally, Kev Abazajian discussed sterile neutrinos, the original dark fermions, and shed light on the tantalizing 3.5 keV line seen from galaxies and clusters of galaxies and its natural explanation in the context of sterile neutrino dark matter.

The following proceedings contributions capture a bit of the flavor of the Simons Symposium on Illuminating Dark Matter. We hope that it will serve as an interesting snapshot of a field in rapid transition, and perhaps that some of the talks presented here will be seen in the future to contain some seeds of insight that ultimately blossomed into the identification of dark matter.

Stony Brook, USA   Rouven Essig
Irvine, USA   Jonathan Feng
Berkeley, USA   Kathryn Zurek
July 2018

Participants

The Simons Symposium on "Illuminating Dark Matter" took place on May 13–19, 2018 at Schloss Elmau, in Elmau, Krün, Germany.

Kevork Abazajian, UC Irvine
Jo Bovy, University of Toronto
Alyson Brooks, Rutgers University
Bernard Carr, Queen Mary, University of London
Aaron Chou, Fermilab
Neal Dalal, Perimeter Institute
Bertrand Echenard, California Institute of Technology
Rouven Essig, Stony Brook University
Jonathan L. Feng, UC Irvine
Roni Harnik, Fermilab
Phil Hopkins, California Institute of Technology
Manoj Kaplinghat, UC Irvine
Alexander Kusenko, UCLA and Kavli IPMU
Rafael F. Lang, Purdue University
Julio Navarro, University of Victoria
Mauro Raggi, Rome La Sapienza
Adam Ritz, University of Victoria
Joshua Ruderman, New York University
Jessie Shelton, University of Illinois at Urbana-Champaign
Javier Tiffenberg, Fermi National Accelerator Laboratory
Tomer Volansky, Tel Aviv University
Neal Weiner, New York University
Kathryn Zurek, Lawrence Berkeley National Laboratory

Contents

Sterile Neutrino/Dark Fermion Dark Matter: Searches in the X-Ray Sky, the Nuclear Physics Laboratory and in Galaxy Formation

Kevork N. Abazajian

Abstract The possibility of dark matter being a particle involved in the generation of neutrino mass has been of interest for over 25 years. Sterile neutrinos—or in the contemporary parlance—dark fermions, are among the simplest and most cited particles which can provide a mechanism for neutrino mass. If one particle of this class has a small mixing, it can be quasi-thermally or nonthermally produced in the early Universe, affect cosmological structure formation, and be detected by X-ray telescopes or laboratory nuclear experiments. A candidate line was detected in 2014, and I review the status of the line and its implications for galaxy formation, proposals for future observations, and laboratory detection.

1 Introduction

One of the most significant discoveries in the past two decades in particle physics was that neutrinos have mass and oscillate between flavor states [1]. The presence of mass requires a mass generation mechanism, and many mechanisms have been proposed [2]. One of the simplest and prevalent mechanisms has been the introduction of Majorana and Dirac type mass terms into the Standard Model Lagrangian:

$$\mathcal{L} \supset -h_{\alpha i} L_\alpha N_i \varphi - \frac{1}{2} M_{ij} N_i N_j + H.c., \tag{1}$$

where $h_{\alpha i}$ are the Yukawa couplings for the flavor states $\alpha = e, \mu, \tau$ and $M_{ij} = M_{ji}$ ($i, j = 1, 2, ...$) are the Majorana masses. The extra particles involved in this mechanism were labeled "sterile neutrinos" due to the lack of their involvement in Standard Model interactions. In a sense, the introduction of these sterile neutrinos was one of the first instances of invoking dark fermion fields to solve a problem in

K. N. Abazajian (✉)
Department of Physics and Astronomy, Center for Cosmology,
University of California, Irvine, CA 92697, USA
e-mail: kevork@uci.edu
URL: http://www-physics.uci.edu/~kevork

© Springer Nature Switzerland AG 2019
R. Essig et al. (eds.), *Illuminating Dark Matter*, Astrophysics and Space
Science Proceedings 56, https://doi.org/10.1007/978-3-030-31593-1_1

1

particle physics. In the contemporary parlance, they would be referred to as "dark fermions" from a "hidden sector" and I interchanged these names below.

This mechanism can be embedded in what is described as the seesaw mechanism, where the smallness of neutrino mass is due to the generation of Majorana masses

$$m_{ij} = \frac{\lambda_{ij} \langle H \rangle^2}{M_N} \sim \frac{m_D^2}{M_N}. \tag{2}$$

For M_N large, $m_{ij} \ll m_D$, providing small neutrino masses, as observed relative to charged lepton masses. The high-scale completion of this mechanism determines the type of seesaw mechanism [3]. The relation in Eq. (1) can be considered simply phenomenologically, and this is dubbed the "new Standard Model" or "neutrino Standard Model" (νSM) [4] or the "neutrino Minimal Standard Model" (νMSM) [5]. In this mechanism, there are only two "heavy" M_N required to produce the observed atmospheric and solar oscillation mass scales. If there is a symmetry in flavor between the "dark sector" fermions (sterile neutrinos) and the Standard Model fermions, we get an extra M_N for free. This dark fermion (sterile neutrino) can have arbitrary mass and mixings, except for where there are constraints on sterile neutrino mixings. In fact, the mixing for this dark fermion (sterile neutrino) can be arbitrarily small, since the lightest mass eigenstate, m_α of the neutrinos may be arbitrarily small

$$\theta \sim \sqrt{\frac{m_\alpha}{M_N}} \ll 1. \tag{3}$$

The arbitrarily small mixing angle is of interest because production and the candidate signals in X-ray indicate mixing angles of order 10^{-7}–10^{-10}. Note that the simplest models of this form cannot accommodate both dark fermions (sterile neutrinos) being dark matter as well as a short baseline neutrino [5]. More complicated models can accommodate both [6].

2 Sterile Neutrino/Dark Fermion Dark Matter: Production and Indirect Detection

It was realized in 1992 that such an extra dark fermion could be dark matter [7], via scattering induced production in mixing with an active neutrino—dubbed the Dodelson-Widrow case. In 1998, this was extended to use a Mikheyev-Smirnov-Wolfenstein resonant production method in universes that could have a nontrivial lepton asymmetry in active neutrinos—dubbed the Shi-Fuller case [8]. Nonresonant Dodelson-Widrow production occurs at temperatures of approximately 100 MeV, where above that temperature the scattering rate to Hubble expansion rate decreases with increasing temperatures as $\Gamma/H \propto T^{-9}$ and below that temperature, the rate of production decreases again as $\Gamma/H \propto T^3$. In the resonant Shi-Fuller case, there can still be some production via the nonresonant production, but the production is enhanced by resonance at higher temperature due to the presence of lepton asymmetry. Therefore, the Shi-Fuller case can produce the requisite total dark matter density with even smaller mixing angles.

In my collaboration with Fuller and Patel, we explored the full parameter space of mixing, mass, and lepton number, where the latter is what connects the two mechanisms of Dodelson-Widrow and Shi-Fuller [9]. (As the lepton number approaches zero, Shi-Fuller becomes Dodelson-Widrow.) We included the effects of the quark-hadron transition that occurs during peak production of the dark matter in much of the parameter space. That work also explored all of the constraints on the dark fermion/sterile neutrino dark matter, including structure formation constraints due to the "warmness" of the dark matter, the cosmic microwave background spectral distortions due to decay, big bang nucleosynthesis, the diffuse X-ray background, and supernova cooling. The strongest constraint was found by Abazajian, Fuller, and Patel to be likely from X-ray emission as observed in relatively local structures like galaxies and clusters of galaxies observed by the contemporary X-ray telescopes *Chandra* and *XMM-Newton*. This is due to the radiative decay of the dark fermion/sterile neutrino, which has no GIM suppression:

$$\Gamma_\gamma(m_s, \sin^2 2\theta) \approx 1.36 \times 10^{-30}\,\mathrm{s}^{-1}\ \left(\frac{\sin^2 2\theta}{10^{-7}}\right) \left(\frac{m_s}{1\,\mathrm{keV}}\right)^5, \qquad (4)$$

for the Majorana case. This decay was first pointed out and calculated by Shrock [10] and independently by Pal & Wolfenstein [11], and for the Majorana case by [12]. The level of constraints from *Virgo Cluster XMM-Newton* observations were explored in Abazajian et al. [13]. In that work, we also explored future sensitivities by the at the time proposed *Constellation-X* mission, as well as pointing out that an equivalent exposure to that large mission "could be obtained by a stacking analysis of the spectra of a number of similar clusters." In the subsequent thirteen years, there was a long history of searches for the line in X-ray data, with no conclusive evidence (for a review, see [14]).

In 2014, Bulbul et al. [15] used a stack of 73 clusters to search for dark matter decay lines, and discovered a ≈5σ line in the sample, in several subsamples, with both detectors aboard *XMM-Newton*. They also saw evidence for the line in *Chandra* observations of the *Perseus* Cluster. This detection was followed up in several searches, with evidence for the line from Andromeda [16], the Milky Way Galactic Center (with *XMM-Newton*) [17], with *SUZAKU* X-ray Space Telescope data toward Perseus [18], in 8 more clusters at >2σ significance [19], and in *NUSTAR* [20] and *Chandra* [21] deep fields' exposure to the Milky Way Galactic Halo, as shown in Fig. 1.

The line was not detected in claimed sensitive observations of stacked galaxies, though the systematic uncertainties in the continuum were of order the signal [22]. It was also not detected in MOS observations of Draco, though the authors of the analysis state the observations do not exclude a dark matter decay interpretation [23].

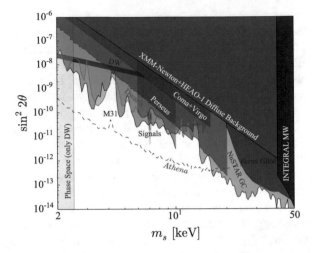

Fig. 1 The full parameter space for sterile neutrino dark matter is shown, for the case where it comprises all of the dark matter. Constraints arise from M31 Horiuchi et al. [24], as well as stacked dwarfs [25]. Also shown are constraints from the diffuse X-ray background [26], and individual clusters "Coma+Virgo" [27]. At higher masses, we show the limits from Fermi GBM [28] and INTEGRAL [29]. The signals near 3.55 keV from M31 and stacked clusters are also shown [15, 16]. The vertical mass constraint only directly applies to the Dodelson-Widrow model being all of the dark matter, labeled "DW," which is now excluded as all of the dark matter. We also show forecast sensitivity of the planned *Athena X-ray Telescope* [30]. This figure is from Ref. [14]

3 Galaxy Formation

There has been considerable interest in this candidate because of the "warmness" of sterile neutrino dark matter, which may alleviate problems in galaxy formation, namely the central density or *Too Big To Fail* problem [32, 33]. This was explored in ranges of the parameter space of the signal, showing how sterile neutrino/dark fermion dark matter could be differentiated from thermal warm dark matter in dark matter only simulations [34, 35], as well as recent full hydrodynamic simulations [36]. It should be emphasized that the fraction of dark matter need not by 100% in order to produce the candidate line but could be as little as 7×10^{-4} in low reheating universe scenarios [14]. In the case of a small fraction of the dark matter being the signal producing sterile neutrino/dark fermion, the dynamics of galaxy formation would be dominated by the predominant form of dark matter, whether it is cold dark matter, self-interacting dark matter, or anything else (Fig. 2).

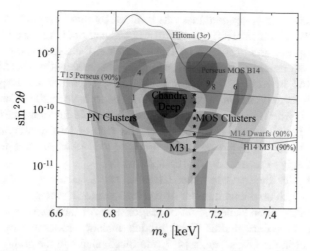

Fig. 2 X-ray line detections consistent with sterile neutrino dark matter are shown here. The dark colored regions are 1, 2 and 3 σ from the MOS (blue) and PN (red) stacked clusters by Bulbul et al. [15], the Bulbul et al. core-removed Perseus cluster (green), and M31 (orange) from Boyarsky et al. [16]. Also shown are the 1 and 2σ regions of the detection in the Galactic Center (GC) [17] as well as the $>2\sigma$ line detections in 1. Abell 85; 2. Abell 2199; 3. Abell 496 (MOS); 4. Abell 496 (PN); 5. Abell 3266; 6. Abell S805; 7. Coma; 8. Abell 2319; 9. Perseus by Iakubovskyi et al. [19]. Numbers in the plot mark the centroid of the regions, with MOS detections in orange and PN in purple. We also show, in purple, the region consistent with the signal in *Chandra* Deep Field observations, with errors given by the flux uncertainty, i.e., not including dark matter profile uncertainties [21]. The lines show constraints at the 90% level from *Chandra* observations of M31 (14) [24], stacked dwarf galaxies (M14) [25], and Suzaku observations of Perseus (T15) [31]. Stars mark models studied in Ref. [32]. This figure is from Ref. [14]

4 The Future

Several follow-up observations and experiments are planned. On the laboratory experiment side, the KArlsruhe TRItium Neutrino (KATRIN) experiment has sensitivities down to mixing angles of $\sim 10^{-8}$ at $m_s \approx 7\,\mathrm{keV}$ [37], and the atom trap experiment Heavy Unseen Neutrinos from Total Energy-momentum Reconstruction (HUNTER) aims to be even more sensitive via energy-momentum reconstruction of K-capture ^{131}Cs [38].

Future searches on the sky include the *Micro-X* and *XQC* sounding rocket experiments, which can be sensitive to the signal parameter space, with campaigns that could occur from the southern hemisphere in the summer of 2019 [39]. Large missions such as *ATHENA* [30] and the *X-ray Surveyor* will be very sensitive to the candidate signal parameter space, but are a decade or more away from launch (2028 launch for *ATHENA* and later for *X-ray Surveyor*). Sooner would be the replacement mission for *Hitomi*, the X-ray Recovery Mission (*XARM*), which is scheduled for launch in March 2021, and would be sensitive to the velocity broadening of the line [15].

A very exciting development that was initiated by this Symposium was the proposal for a CUBESAT mission by a team at Fermi National Accelerator Laboratory led by S. Timpone, and presented by Roni Harnik at the Symposium. The mission would use newly designed CCDs from Fermilab that are sensitive to 3.5 keV photons on a CUBESAT in order to search for the line from exposure to a large fraction of the sky. Such an experiment is roughly estimated to have $\sim 20\sigma$ sensitivity to the 3.5 keV signal, with a mere 90 min exposure. I was able to connect the instrumentalist team at Fermilab with space mission specialists at Jet Propulsion Lab, led by Olivier Doré, to help draft a more robust proposal to NASA for funding what would be an exceedingly high-impact CUBESAT mission.

On the physics theory side, Alex Kusenko and I started a new project at the Symposium which studies a number of production mechanisms besides Dodelson-Widrow and Shi-Fuller, which could be responsible for the decay line and be all of or a fraction of the dark matter. These models include production by the decay of a gauge singlet in the Higgs sector [40], and mechanisms in the split seesaw model [41]. This collaboration is ongoing thanks to the Symposium.

Acknowledgements Immense thanks goes to the Simons Foundation for hosting this excellent Symposium, where many new directions of dark matter physics and astrophysics were explored. Several new collaborations and research connections resulted. The work presented here was supported in part by NSF Grant PHY-1620638.

References

1. C. Patrignani et al., Chin. Phys. C **40**(10), 100001 (2016). https://doi.org/10.1088/1674-1137/40/10/100001
2. A. de Gouvêa, Ann. Rev. Nucl. Part. Sci. **66**, 197 (2016). https://doi.org/10.1146/annurev-nucl-102115-044600
3. R.N. Mohapatra, A.Y. Smirnov, Ann. Rev. Nucl. Part. Sci. **56**, 569 (2006). https://doi.org/10.1146/annurev.nucl.56.080805.140534
4. A. de Gouvea, Phys. Rev. D **72**, 033005 (2005). https://doi.org/10.1103/PhysRevD.72.033005
5. T. Asaka, S. Blanchet, M. Shaposhnikov, Phys. Lett. B **631**, 151 (2005). https://doi.org/10.1016/j.physletb.2005.09.070
6. D. Borah, Phys. Rev. D **94**(7), 075024 (2016). https://doi.org/10.1103/PhysRevD.94.075024
7. S. Dodelson, L.M. Widrow, Phys. Rev. Lett. **72**, 17 (1994). https://doi.org/10.1103/PhysRevLett.72.17
8. X.D. Shi, G.M. Fuller, Phys. Rev. Lett. **82**, 2832 (1999). https://doi.org/10.1103/PhysRevLett.82.2832
9. K. Abazajian, G.M. Fuller, M. Patel, Phys. Rev. D **64**, 023501 (2001). https://doi.org/10.1103/PhysRevD.64.023501
10. R. Shrock, Phys. Rev. D **9**, 743 (1974). https://doi.org/10.1103/PhysRevD.9.743
11. P.B. Pal, L. Wolfenstein, Phys. Rev. D **25**, 766 (1982). https://doi.org/10.1103/PhysRevD.25.766
12. V.D. Barger, R.J.N. Phillips, S. Sarkar, Phys. Lett. B **352**, 365 (1995). https://doi.org/10.1016/0370-2693(95)00486-5, https://doi.org/10.1016/0370-2693(95)00831-5. [Erratum: Phys. Lett. B356,617(1995)]
13. K. Abazajian, G.M. Fuller, W.H. Tucker, Astrophys. J. **562**, 593 (2001). https://doi.org/10.1086/323867

14. K.N. Abazajian, Phys. Rept. **711–712**, 1 (2017). https://doi.org/10.1016/j.physrep.2017.10.003

15. E. Bulbul, M. Markevitch, A. Foster, R.K. Smith, M. Loewenstein, S.W. Randall, Astrophys. J. **789**, 13 (2014). https://doi.org/10.1088/0004-637X/789/1/13

16. A. Boyarsky, O. Ruchayskiy, D. Iakubovskyi, J. Franse, Phys. Rev. Lett. **113**, 251301 (2014). https://doi.org/10.1103/PhysRevLett.113.251301

17. A. Boyarsky, J. Franse, D. Iakubovskyi, O. Ruchayskiy, Phys. Rev. Lett. **115**, 161301 (2015). https://doi.org/10.1103/PhysRevLett.115.161301

18. O. Urban, N. Werner, S.W. Allen, A. Simionescu, J.S. Kaastra, L.E. Strigari, Mon. Not. Roy. Astron. Soc. **451**(3), 2447 (2015). https://doi.org/10.1093/mnras/stv1142

19. D. Iakubovskyi, E. Bulbul, A.R. Foster, D. Savchenko, V. Sadova, preprint, arXiv:1508.05186 [astro-ph.HE] (2015)

20. A. Neronov, D. Malyshev, D. Eckert, Phys. Rev. D **94**(12), 123504 (2016). https://doi.org/10.1103/PhysRevD.94.123504

21. N. Cappelluti, E. Bulbul, A. Foster, P. Natarajan, M.C. Urry, M.W. Bautz, F. Civano, E. Miller, R.K. Smith, Astrophys. J. **854**(2), 179 (2018). https://doi.org/10.3847/1538-4357/aaaa68

22. M.E. Anderson, E. Churazov, J.N. Bregman, Mon. Not. Roy. Astron. Soc. **452**(4), 3905 (2015). https://doi.org/10.1093/mnras/stv1559

23. O. Ruchayskiy, A. Boyarsky, D. Iakubovskyi, E. Bulbul, D. Eckert, J. Franse, D. Malyshev, M. Markevitch, A. Neronov, Mon. Not. Roy. Astron. Soc. **460**(2), 1390 (2016). https://doi.org/10.1093/mnras/stw1026

24. S. Horiuchi, P.J. Humphrey, J. Onorbe, K.N. Abazajian, M. Kaplinghat, S. Garrison-Kimmel, Phys. Rev. D **89**(2), 025017 (2014). https://doi.org/10.1103/PhysRevD.89.025017

25. D. Malyshev, A. Neronov, D. Eckert, Phys. Rev. D **90**, 103506 (2014). https://doi.org/10.1103/PhysRevD.90.103506

26. A..Boyarsky, A. Neronov, O. Ruchayskiy, M. Shaposhnikov, Mon. Not. Roy. Astron. Soc. **370**, 213 (2006). https://doi.org/10.1111/j.1365-2966.2006.10458.x

27. A. Boyarsky, A. Neronov, O. Ruchayskiy, M. Shaposhnikov, Phys. Rev. D **74**, 103506 (2006). https://doi.org/10.1103/PhysRevD.74.103506

28. K.C.Y. Ng, S. Horiuchi, J.M. Gaskins, M. Smith, R. Preece, Phys. Rev. D **92**(4), 043503 (2015). https://doi.org/10.1103/PhysRevD.92.043503

29. A. Boyarsky, D. Malyshev, A. Neronov, O. Ruchayskiy, Mon. Not. Roy. Astron. Soc. **387**, 1345 (2008). https://doi.org/10.1111/j.1365-2966.2008.13003.x

30. A. Neronov, D. Malyshev, Phys. Rev. D **93**(6), 063518 (2016). https://doi.org/10.1103/PhysRevD.93.063518

31. T. Tamura, R. Iizuka, Y. Maeda, K. Mitsuda, N.Y. Yamasaki, Publ. Astron. Soc. Jpn. **67**, 23 (2015). https://doi.org/10.1093/pasj/psu156

32. T. Venumadhav, F.Y. Cyr-Racine, K.N. Abazajian, C.M. Hirata, Phys. Rev. D **94**(4), 043515 (2016). https://doi.org/10.1103/PhysRevD.94.043515

33. K.N. Abazajian, Phys. Rev. Lett. **112**(16), 161303 (2014). https://doi.org/10.1103/PhysRevLett.112.161303

34. S. Horiuchi, B. Bozek, K.N. Abazajian, M. Boylan-Kolchin, J.S. Bullock, S. Garrison-Kimmel, J. Onorbe, Mon. Not. Roy. Astron. Soc. **456**(4), 4346 (2016). https://doi.org/10.1093/mnras/stv2922

35. B. Bozek, M. Boylan-Kolchin, S. Horiuchi, S. Garrison-Kimmel, K. Abazajian, J.S. Bullock, Mon. Not. Roy. Astron. Soc. **459**(2), 1489 (2016). https://doi.org/10.1093/mnras/stw688

36. B. Bozek, et al., Mon. Not. Roy. Astron. Soc. **483**(3), 4086 (2019). https://doi.org/10.1093/mnras/sty3300

37. N.M.N. Steinbrink, J.D. Behrens, S. Mertens, P.C.O. Ranitzsch, C. Weinheimer, Eur. Phys. J. C **78**(3), 212 (2018). https://doi.org/10.1140/epjc/s10052-018-5656-9

38. P.F. Smith, New J. Phys. **21**(5), 053022 (2019). https://doi.org/10.1088/1367-2630/ab1502

39. E. Figueroa-Feliciano et al., Astrophys. J. **814**(1), 82 (2015). https://doi.org/10.1088/0004-637X/814/1/82

40. K. Petraki, A. Kusenko, Phys. Rev. D **77**, 065014 (2008). https://doi.org/10.1103/PhysRevD. 77.065014
41. A. Kusenko, F. Takahashi, T.T. Yanagida, Phys. Lett. B **693**, 144 (2010). https://doi.org/10. 1016/j.physletb.2010.08.031

Constraining the Small-Scale Clustering of Dark Matter with Stellar Streams

Jo Bovy

Abstract The degree of dark matter clustering on small scales presents a strong constraint on its physical nature. One of the most promising avenues for determining the clustering of dark matter on the smallest scales employs narrow stellar streams in the halo of the Milky Way. In this contribution, I review recent progress in modeling the effect of dark matter substructure on the structure of stellar streams and recent constraints on the amount of small ($\approx 10^7 \, M_\odot$) dark matter substructure in the inner Milky Way halo. The next few years will likely see a large amount of progress both in the modeling of stellar streams and in the quantity and quality of the available data and I discuss future challenges and opportunities in this area.

1 Introduction

Many decades after the discovery of dark matter through its effect on galactic velocities, we still have few clues as to what its mass and interactions with itself and standard model particles are. The prevailing dark matter paradigm posits that it is a particle that weakly, if at all, interacts with itself and ordinary matter—it is "collisionless"—and that has been nonrelativistic throughout cosmological structure formation—it is "cold". This cold, collisionless paradigm known as "CDM" (cold dark matter) is consistent with all observations, notwithstanding some controversies on the structure of dwarf galaxies that may be resolved by interesting dark matter physics, but do not require it (see contributions by Brooks, Hopkins, and Navarro in this volume).

If dark matter is cold and collisionless, galaxy formation simulations demonstrate that it should cluster strongly on scales far below galactic scales and this clustering manifests itself as a large abundance of gravitationally bound dark matter *subhalos* orbiting within the dark matter halos of galaxies like our own Milky Way [1–4]. However, for many well-motivated models of dark matter this clustering is reduced or entirely removed, leading to a reduced or entirely absent populations of subhalos.

J. Bovy (✉)
Department of Astronomy and Astrophysics, University of Toronto,
50 St. George Street, Toronto, ON M5S 3H4, Canada
e-mail: bovy@astro.utoronto.ca

© Springer Nature Switzerland AG 2019
R. Essig et al. (eds.), *Illuminating Dark Matter*, Astrophysics and Space
Science Proceedings 56, https://doi.org/10.1007/978-3-030-31593-1_2

This is the case, for example, for ultralight axion dark matter [5] or sterile neutrinos [6, 7]. While the abundance of dark matter subhalos can be traced at high mass by the abundance of dwarf galaxies, the mapping between dark matter subhalos and galaxies are highly uncertain and strongly dependent on poorly understood baryonic effects on the abundance of dwarf galaxies.

In the absence of any other known dark matter interactions, constraining the amount of dark matter substructure in galactic halos requires observing it through its gravitational effect. The two most promising methods for doing this currently are (a) gravitational lensing [8–12] and (b) stellar streams. Stellar streams are narrow or "cold"—$\approx 1\%$ width in position and velocity compared to the smooth stellar halo—stellar features that are formed through the tidal stripping of a globular cluster of stars or of a small dwarf galaxy (commonly referred to as the "progenitor" of the stream). Over many pericentric passages, the progenitor loses stars at a small rate through the stronger tides near the galactic center and these stars have slightly higher or lower orbital energies than the progenitor. Due to this energy difference, the stream stars slowly drift away from the progenitor over time along a path that is close to the orbit of the progenitor. Over many orbits, this produces an approximately constant density stream of stars to emerge along with its past and future orbit. The poster child of this process is the stellar stream emanating from the Pal 5 globular cluster [13], which has now been detected to extend out over about 30° [14]. Because of the small range of orbital energies within a stream, a stellar stream would remain narrow for much longer than a Hubble time in the absence of any perturbations.

2   Stellar Streams and Dark Matter

2.1   Simple Considerations

A passing dark matter subhalo can gravitationally perturb the orbits of the stars in a stream if it passes closely enough. In the impulse approximation, a subhalo of mass M that passes at a relative velocity v with the closest approach at impact parameter b induces a velocity kick Δv

$$\Delta v = \frac{2GM}{bv} = 1.3\,\mathrm{km\,s^{-1}} \left(\frac{M}{10^7\,M_\odot}\right) \left(\frac{b}{0.3\,\mathrm{kpc}}\right)^{-1} \left(\frac{v}{220\,\mathrm{km\,s^{-1}}}\right)^{-1}. \quad (1)$$

To determine the masses of dark matter subhalos that stream perturbations are sensitive to, we can compare this Δv to the internal velocity dispersion within the stream σ_v. Assuming that the typical $v \approx V_c$ with V_c the circular velocity, that the impact parameter $b = r_s$ (necessary to obtain a sizeable signal), with r_s the scale parameter of the subhalo, and using the relation between mass and scale radius from numerical simulations $r_s \approx 1\,\mathrm{kpc}\,\left(M/10^8\,M_\odot\right)^{0.5}$ [15], we find that for

$$M \gtrsim 3 \times 10^7 \, M_\odot \left(\frac{\sigma_v}{2 \, \text{km s}^{-1}} \right)^2 \left(\frac{v}{220 \, \text{km s}^{-1}} \right)^2 , \tag{2}$$

where we have used $v \approx V_c = 220 \, \text{km s}^{-1}$ (the Milky Way's circular velocity), a single impact produces a velocity kick larger than the internal velocity dispersion in the stream. Thus, the effect of $\mathcal{O}(10^7 \, M_\odot)$ dark matter subhalos on stellar streams is *large*.

Given enough stars in the stream, we can measure deviations much better than the internal dispersion. It is clear from Eq. (1) that the scale over which the impulse varies is $\approx b$. Thus, we need enough stars over a scale b to see the velocity impact from a subhalo of mass M. Assuming we observe N stars deg^{-1} along the stream (collapsing the stream along all but the longest direction on the sky), we can compute that the velocity impact Δv from a single impact is observable when

$$\frac{2GM}{bv} > \frac{\sigma_v}{\sqrt{N \frac{180}{\pi} \frac{b}{D}}} , \tag{3}$$

or with the same assumptions as above

$$M \gtrsim 3 \times 10^6 \, M_\odot \left(\frac{\sigma_v}{2 \, \text{km s}^{-1}} \right)^{4/3} \left(\frac{v}{220 \, \text{km s}^{-1}} \right)^{4/3} \left(\frac{N}{25 \, \text{deg}^{-1}} \right)^{-2/3} \tag{4}$$
$$\times \left(\frac{D}{20 \, \text{kpc}} \right)^{2/3} \quad \textbf{(single impact, velocity)} ,$$

where we use a distance and stellar density that roughly correspond to current observations of the Pal 5 stream.

The perturbation to the velocities of stars within the stream cause the stars to move away from the point of impact and create an underdensity or "gap" in the stream. In the absence of any velocity dispersion within the stream, impacts of any mass create an $\mathcal{O}(1)$ density perturbation after a time $\Delta t \approx b/\Delta v = b^2 v/(2GM)$, which is approximately independent of M (because $b^2 \approx r_s^2 \propto M$ for subhalos). However, velocity dispersion means that the gap has only

$$\Delta t \approx \frac{b}{\sigma_v} = 0.15 \, \text{Gyr} \left(\frac{M}{10^7 \, M_\odot} \right)^{1/2} \left(\frac{\sigma_v}{2 \, \text{km s}^{-1}} \right) , \tag{5}$$

to grow before it starts to fill in again. The maximum density perturbation is then (see, e.g., [16])

$$\frac{\Delta \rho}{\rho} \approx \frac{\Delta v}{\sigma_v} \approx 0.5 \left(\frac{M}{10^7 \, M_\odot} \right)^{1/2} \left(\frac{\sigma_v}{2 \, \text{km s}^{-1}} \right)^{-1} \left(\frac{v}{220 \, \text{km s}^{-1}} \right)^{-1} . \tag{6}$$

The density perturbation extends over a physical scale of $\Delta x \approx \Delta v \Delta t$ or

$$\Delta x \approx 0.2 \, \text{kpc} \left(\frac{M}{10^7 \, M_\odot} \right) \left(\frac{\sigma_v}{2 \, \text{km s}^{-1}} \right)^{-1} \left(\frac{v}{220 \, \text{km s}^{-1}} \right)^{-1}. \tag{7}$$

Assuming that the uncertainty in the measured density along a stream is given by the Poisson uncertainty in the number counts, this density contrast is observable for

$$M \gtrsim 5 \times 10^6 \, M_\odot \left(\frac{\sigma_v}{2 \, \text{km s}^{-1}} \right)^{3/2} \left(\frac{v}{220 \, \text{km s}^{-1}} \right)^{3/2} \left(\frac{N}{25 \, \text{deg}^{-1}} \right)^{-1/2} \left(\frac{D}{20 \, \text{kpc}} \right)^{1/2} \tag{8}$$

$$\textbf{(single impact, density)}.$$

These simple estimates have serious limitations in that they do not fully take into account how long the signal is observable or how many impacts we expect with $b \approx r_s$ or their velocity distribution. The velocity signal is largest right after the impact and then starts to decay away on the same timescale as in Eq. (5) due to phase-mixing in the stream. In this sense, the estimate in Eq. (4) represents the most optimistic case where the impact is observed right after it has occurred. The density estimate in Eq. (8) takes some of the limitations caused by phase-mixing into account and is, therefore, a more realistic estimate. More impacts occur for the more abundant subhalos with lower masses, which also means that they are more likely to happen with a smaller impact velocity v, giving rise to a larger signal. Despite these limitations, these simple considerations show that cold stellar streams are uniquely sensitive to small dark matter subhalos and they provide a sense of the relevant mass scales and other ingredients.

2.2   Results from Detailed Simulations

To fully characterize the potential sensitivity of stellar streams to a CDM-like populations of dark matter subhalos orbiting within a galaxy like the Milky Way, it is necessary to perform simulations. Until recently, this was done using expensive N-body simulations [17–20] and only dozens of different realizations were studied. Reference [15] developed a much faster method for computing the effects of subhalo perturbations on stellar streams fully taking into account the effect of phase-mixing and the noncircularity of the stream's orbit, by building on the simple model for the dynamics of stellar streams from Refs. [21, 22]. This has allowed thousands of simulations to be performed and the effect of different mass ranges, subhalo properties, stream age, subhalo velocity distributions, etc. could be studied. In particular, Ref. [15] proposed using a power spectrum approach to characterize the observed and computed density and stream position perturbations, because the power on different scales is indicative of the presence of subhalos with different masses. Figure 1 shows example power spectra expected from a CDM-like population of subhalos in different mass ranges impacting a stream like the observed GD-1 stream [23].

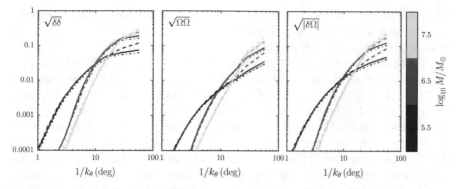

Fig. 1 Median power spectra of the fluctuations in the density (left panel) and the mean track stream track—a proxy for velocity variations along the stream—relative to those in the unperturbed stream (middle panel) as well as the density-track cross-correlation (right panel) for impacts of different masses for a GD-1-like stream. Different mass ranges give rise to power on different scales; this effect can be used to determine the number of dark matter subhalos with different masses. Fluctuations in the density and the stream location are strongly correlated. Figure from Ref. [15]

These simulations show that perturbations to the stream density are more sensitive than perturbations to the velocities or positions of stars in the stream. Thus, detailed measurements of the density of multiple streams are the most promising avenue for constraining the dark matter subhalo population in the Milky Way. Cross-correlations between the density and the velocity or spatial location of the stream are the next most sensitive method for detecting subhalo perturbations; these would provide an important check on the validity of any claimed detection, as cross-correlations between density and velocity is much less sensitive to observational effects such as dust extinction, survey selection biases, etc. than the density.

These simulations also reveal that the small effect from very low mass dark matter subhalos with masses $M \approx 10^5 \, M_\odot$—which produce $\approx 5\%$ density perturbations (see Eq. (6))—are observable in the aggregate when $N \gtrsim 200 \, \mathrm{deg}^{-1}$ (which makes sense according to Eq. (8)). This means that a cutoff in the mass function of dark matter subhalos caused, for example, by free streaming of warm dark matter should be detectable in the near future at $M \gtrsim 10^6 \, M_\odot$.

3 Current Constraints

Stellar streams are very low surface brightness stellar overdensities that require sophisticated filters applied to the observed positions, velocities, colors, and magnitudes to be detected. While more than a dozen narrow stellar streams are currently known, only a single one, the Pal 5 stream, has been observed in sufficient detail to measure perturbations in its density at the level expected from subhalo perturbations.

The Pal 5 stream is an attractive target for subhalo searches because unlike many of the other narrow stellar streams, its progenitor cluster is known. This simplifies the modeling of this system significantly.

Reference [15] used the Pal 5 density measurements obtained using deep color-magnitude diagrams along the stream by Ref. [14] to compute the power spectrum of density fluctuations along the trailing stream. This power spectrum is displayed in Fig. 2. On all but the largest scales, the power in the density is dominated by the random uncertainties due to the low number counts in the stream. The colored curves show the expected power spectrum due to subhalo perturbations to the stream for different normalizations of the mass function relative to that expected in CDM. It is clear that we do not expect to detect any perturbations on small scales with the current quality of data, but on the largest scales, the observed power is roughly consistent with expectations.

Reference [15] performed a more sophisticated analysis using Approximate Bayesian Computation to match the simulated density power spectra to the observed power spectrum and determine a posterior distribution function (PDF) for the number of subhalos in the inner Milky Way relative to CDM expectations. The PDFs obtained assuming different mass ranges of perturbations are displayed in Fig. 2. It is clear that the current density measurements are sensitive to masses as small as $M \approx 3 \times 10^6 \ M_\odot$ and that the constraints on the number of subhalos are fully consistent with the CDM expectation (zero on the logarithmic scale in Fig. 2). However,

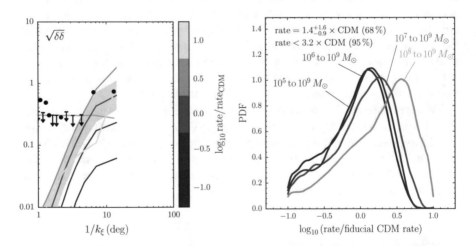

Fig. 2 Observed density power spectrum for the Pal 5 stream (black points, left panel) and simulated power spectra for a CDM-like population of dark matter subhalos in the inner Milky Way scaled by a constant (colored curves, left panel). With current density uncertainties, only the power on the largest scales is observationally accessible. The posterior distribution function for the number of subhalos in the inner Milky Way compared to that expected for a CDM-like population. This constraint is obtained by matching the observed density power spectrum with an Approximate Bayesian Computation technique. The current constraints are fully consistent with the expected number of subhalos in CDM. Figure from Ref. [15]

this analysis ignores the potential effect from other perturbers such as the bar, spiral structure, and Giant Molecular Clouds (GMCs) and is, therefore, a more robust upper limit than a measurement. Further work investigating the effect of these baryonic perturbers is required for a more definitive measurement using Pal 5.

4 Future Challenges and Opportunities

It is clear from the discussion above that stellar streams present an enormous opportunity for constraining the small-scale structure of the Milky Way's dark matter halo and, thus, for providing important constraints on the physical nature of dark matter. Currently, new data on stellar streams is emerging from the *Gaia* survey [24], which will allow new and existing nearby stellar streams ($D \lesssim 15\,\mathrm{kpc}$) to be mapped in much more detail than we have today. However, while this will lead to a better understanding of the structure of stellar streams and to better constraints on the smooth component of the dark matter halo [25], unless many new nearby streams are uncovered by *Gaia*, this will not lead to a substantial improvement in the constraints on dark matter subhalos.

Further progress in this field will come from the LSST survey, scheduled to start science operations at the end of 2022, and surveys such as WFIRST and CASTOR. These surveys will provide deep photometry of objects over wide regions of the sky, with photometric precision good enough to create almost background free maps of stellar streams far below their main sequence turn off (leading to $N \approx \mathcal{O}(100)\,\mathrm{deg}^{-1}$ in the parlance of Sect. 2.1). These will be good enough to bear out the projections from Ref. [15], that the subhalo mass function can be constrained down to $M \approx 10^5\,M_{\odot}$.

In the LSST era, the opportunities when using stellar streams to constrain the small-scale structure of dark matter are

- Stellar streams are sensitive to the dark matter subhalo abundance localized in the three-dimensional region covered by the stream's orbit. Thus, stellar streams at different Galactocentric distances can determine the radial behavior of the subhalo mass function.
- Stellar streams are sensitive to the subhalo abundance over a large fraction of their lifetimes, therefore, they could potentially constrain the evolution of the subhalo abundance over time.
- Stellar streams are also sensitive to partially disrupted dark matter subhalos [26] and could thus constrain the population of tidally disrupting subhalos.
- Detailed density, position, and velocity measurements of stellar streams can fully characterize the mass, internal profile, and fly-by velocity for impacts with larger dark matter subhalos ($M \gtrsim 10^7\,M_{\odot}$) [27]; for recent fly-bys this could lead to robust predictions for the present location of dark subhalos, which would make

excellent targets for gamma-ray and similar searches for dark matter annihilation signals.

To fully realize the potential of stellar streams, a few challenges still need to be overcome:

- The influence of baryonic perturbers: for a stream such as Pal 5, which is on a prograde—that is, in the same direction of the Galactic disk's rotation—orbit around the Galaxy, the effect of the bar [28] and molecular clouds [29] may play a large role. Further simulations of the effect of the bar and molecular clouds are required to characterize the likely effect of these perturbers on the density and stream position of Pal 5 and other streams. Streams such as GD-1, which is on a retrograde orbit, likely feel little to no effect from these perturbers and are in this sense better targets for subhalo searches.
- The effect of uncertainty in the Galactic potential: stream models require a model for the Galactic potential. While the location of the stream itself is a strong constraint on the Galactic potential (e.g., Ref. [25]), the effect of errors in the Galactic potential model (including those from large perturbers such as the Large Magellanic Cloud) needs to be better understood.
- Predictions in the face of uncertainty in the stream model: For many cold streams, the progenitor is unknown, which causes uncertainty in the predictions for the density structure, because the effects of episodic stripping over a finite time lead to both small and large-scale density variations at different distances from the progenitor. Little attention has been paid so far to the importance of this uncertainty. Similarly, the internal structure and dynamics of the progenitor have been largely ignored, even though globular clusters display an interesting array of dynamical effects due to their high concentrations and short dynamical times. This could affect the distribution of stellar masses along the stream (e.g., Ref. [30]) and thus the observed stellar density.
- Uncertainty in the predicted population of subhalos: While the results from dark matter-only N-body simulations of the formation of a Milky Way-like galaxy give detailed predictions for the subhalo mass function, the massive baryonic disk that sits near the center of the Milky Way tidally disrupts a fraction of the subhalos. The predictions from simulations that include a disk (or full hydrodynamics) seem to indicate a depletion by a factor of two to four (e.g., Refs. [31–33]), but different simulations do not agree and the depletion factor is a strong function of the radius of the subhalo. *Measurements* of the subhalo abundance in the Milky Way with streams at different radii would provide a strong test of these simulations.

5   Conclusion

Stellar streams are one of the best ways to test one of the basic predictions of the cold, collisionless dark matter paradigm: that abundant small-scale clustering of dark matter occurs and leads to a population of low mass, completely dark subhalos in the halos of galaxies. In conjunction with complementary measurements from gravitational lensing, stellar streams are in a strong position to robustly test this prediction within the next five years. Because stellar streams (and gravitational lensing) only depend on the gravitational interactions between dark matter and ordinary matter, they perform well in the "nightmare" scenario where no nongravitational dark matter interactions can be detected in the laboratory or astrophysically.

Acknowledgements Many thanks to the organizers of this workshop and to the Simons Foundation for providing a stimulating environment for discussions. This work was supported by the Natural Sciences and Engineering Research Council of Canada (NSERC; funding reference number RGPIN-2015-05235) and by an Alfred P. Sloan Fellowship.

References

1. A. Klypin, A.V. Kravtsov, O. Valenzuela, F. Prada, ApJ **522**, 82 (1999). https://doi.org/10.1086/307643
2. B. Moore, S. Ghigna, F. Governato, G. Lake, T. Quinn, J. Stadel, P. Tozzi, ApJ **524**, L19 (1999). https://doi.org/10.1086/312287
3. V. Springel, J. Wang, M. Vogelsberger, A. Ludlow, A. Jenkins, A. Helmi, J.F. Navarro, C.S. Frenk, S.D.M. White, MNRAS **391**, 1685 (2008). https://doi.org/10.1111/j.1365-2966.2008.14066.x
4. J. Diemand, M. Kuhlen, P. Madau, M. Zemp, B. Moore, D. Potter, J. Stadel, Nature **454**, 735 (2008). https://doi.org/10.1038/nature07153
5. L. Hui, J.P. Ostriker, S. Tremaine, E. Witten, Phys. Rev. D **95**(4), 043541 (2017). https://doi.org/10.1103/PhysRevD.95.043541
6. S. Dodelson, L.M. Widrow, Phys. Rev. Lett. **72**, 17 (1994). https://doi.org/10.1103/PhysRevLett.72.17
7. X. Shi, G.M. Fuller, Phys. Rev. Lett. **82**, 2832 (1999). https://doi.org/10.1103/PhysRevLett.82.2832
8. R.B. Metcalf, P. Madau, ApJ **563**, 9 (2001). https://doi.org/10.1086/323695
9. N. Dalal, C.S. Kochanek, ApJ **572**, 25 (2002). https://doi.org/10.1086/340303
10. S. Vegetti, D.J. Lagattuta, J.P. McKean, M.W. Auger, C.D. Fassnacht, L.V.E. Koopmans, Nature **481**, 341 (2012). https://doi.org/10.1038/nature10669
11. Y. Hezaveh, N. Dalal, G. Holder, T. Kisner, M. Kuhlen, L. Perreault Levasseur, J. Cosmol. Astropart. Phys. **11**, 048 (2016). https://doi.org/10.1088/1475-7516/2016/11/048
12. F.Y. Cyr-Racine, C.R. Keeton, L.A. Moustakas, arXiv (2018)
13. M. Odenkirchen, E.K. Grebel, C.M. Rockosi et al., ApJ **548**, L165 (2001). https://doi.org/10.1086/319095
14. R.A. Ibata, G.F. Lewis, N.F. Martin, ApJ **819**, 1 (2016). https://doi.org/10.3847/0004-637X/819/1/1
15. J. Bovy, D. Erkal, J.L. Sanders, MNRAS **466**, 628 (2017). https://doi.org/10.1093/mnras/stw3067
16. D. Erkal, V. Belokurov, MNRAS **450**, 1136 (2015). https://doi.org/10.1093/mnras/stv655
17. K.V. Johnston, D.N. Spergel, C. Haydn, ApJ **570**, 656 (2002). https://doi.org/10.1086/339791
18. R.A. Ibata, G.F. Lewis, M.J. Irwin, T. Quinn, MNRAS **332**, 915 (2002). https://doi.org/10.1046/j.1365-8711.2002.05358.x
19. R.G. Carlberg, ApJ **705**, L223 (2009). https://doi.org/10.1088/0004-637X/705/2/L223
20. R.G. Carlberg, ApJ **748**, 20 (2012). https://doi.org/10.1088/0004-637X/748/1/20
21. J. Bovy, ApJ **795**, 95 (2014). https://doi.org/10.1088/0004-637X/795/1/95
22. J. Bovy, ApJS **216**, 29 (2015). https://doi.org/10.1088/0067-0049/216/2/29

23. C.J. Grillmair, O. Dionatos, ApJ **643**, L17 (2006). https://doi.org/10.1086/505111
24. Gaia Collaboration, T. Prusti, J.H.J. de Bruijne, A.G.A. Brown, A. Vallenari, C. Babusiaux, C.A.L. Bailer-Jones, U. Bastian, M. Biermann, D.W. Evans, et al., A&A **595**, A1 (2016). https://doi.org/10.1051/0004-6361/201629272
25. J. Bovy, A. Bahmanyar, T.K. Fritz, N.a. Kallivayalil, GCIP dynamics. Galaxy: struct. **833**, 31 (2016). https://doi.org/10.3847/1538-4357/833/1/31
26. J. Bovy, Phys. Rev. Lett. **116**(12), 121301 (2016). https://doi.org/10.1103/PhysRevLett.116.121301
27. D. Erkal, V. Belokurov, MNRAS **454**, 3542 (2015). https://doi.org/10.1093/mnras/stv2122
28. S. Pearson, A.M. Price-Whelan, K.V. Johnston, Nat. Astron. **1**, 633 (2017). https://doi.org/10.1038/s41550-017-0220-3
29. N.C. Amorisco, F.A. Gómez, S. Vegetti, S.D.M. White, MNRAS **463**, L17 (2016). https://doi.org/10.1093/mnrasl/slw148
30. E. Balbinot, M. Gieles, MNRAS **474**, 2479 (2018). https://doi.org/10.1093/mnras/stx2708
31. E. D'Onghia, V. Springel, L. Hernquist, D. Keres, ApJ **709**, 1138 (2010). https://doi.org/10.1088/0004-637X/709/2/1138
32. T. Sawala, P. Pihajoki, P.H. Johansson, C.S. Frenk, J.F. Navarro, K.A. Oman, S.D.M. White, MNRAS **467**, 4383 (2017). https://doi.org/10.1093/mnras/stx360
33. S. Garrison-Kimmel, A. Wetzel, J.S. Bullock, P.F. Hopkins, M. Boylan-Kolchin, C.A. Faucher-Giguère, D. Kereš, E. Quataert, R.E. Sanderson, A.S. Graus, T. Kelley, MNRAS **471**, 1709 (2017). https://doi.org/10.1093/mnras/stx1710

Understanding Dwarf Galaxies in Order to Understand Dark Matter

Alyson M. Brooks

Abstract Much progress has been made in recent years by the galaxy simulation community in making realistic galaxies, mostly by more accurately capturing the effects of baryons on the structural evolution of dark matter halos at high resolutions. This progress has altered theoretical expectations for galaxy evolution within a Cold Dark Matter (CDM) model, reconciling many earlier discrepancies between theory and observations. Despite this reconciliation, CDM may not be an accurate model for our Universe. Much more work must be done to understand the predictions for galaxy formation within alternative dark matter models.

1 Introduction: The Need to Understand Baryons to Understand Dark Matter

Most of the matter in our Universe resides in an unknown component that we refer to as "dark matter." There is six times more mass in dark matter than ordinary matter, which astronomers refer to as baryons. The large-scale distribution of galaxies suggests that dark matter that is "cold" (because it travels slowly compared to the speed of light) provides an excellent description of our Universe [1]. However, when astronomers observe galaxies they are viewing only the ordinary matter that emits and absorbs photons.

Everything that we have learned about dark matter we have learned from astrophysics.[1] The dark matter structure of galaxies is currently the primary method used to constrain the properties of dark matter.

---

[1] With one exception: dark matter direct detection experiments have ruled out a parameter space of cross sections for interactions between dark matter and baryons.

A. M. Brooks (✉)
Department of Physics and Astronomy, Rutgers, the State University of New Jersey,
136 Frelinghuysen Rd., Piscataway, NJ 08854, USA
e-mail: abrooks@physics.rutgers.edu
URL: http://physics.rutgers.edu/~abrooks/index.html

© Springer Nature Switzerland AG 2019
R. Essig et al. (eds.), *Illuminating Dark Matter*, Astrophysics and Space
Science Proceedings 56, https://doi.org/10.1007/978-3-030-31593-1_3

For decades it has been assumed that, because there is so much more dark matter than ordinary matter, dark matter dominates the gravity in the Universe, and that wherever the dark matter is most dense, gas and stars must be there. This assumption led theorists to make predictions for the formation of galaxies that either entirely neglected or poorly modeled the physics of gas and stars. In doing so, a number of discrepancies between galaxy formation theory and observations were identified, particularly on "small scales," i.e., in small galaxies and in the central regions of galaxies. These discrepancies have evaded solution for so many years that they have become known collectively as the "small scale crisis" of the Cold Dark Matter (CDM) model for galaxy formation.

However, in the last few years, there has been a paradigm shift, in which many astronomers now recognize the importance of including baryonic physics to solve CDM's small-scale problems. Two of the most critical problems have been the "cusp-core problem" and the "missing satellites problem." Both problems are generally now agreed to be alleviated (or even solved) by the inclusion of baryonic physics.

Many simulators have demonstrated that energy injection from stars (usually referred to as "feedback") in the form of both supernovae and energy from young, massive stars (i.e., ionization, radiation pressure, and momentum injection from winds) can push the dark matter out of the central $\sim$kpc of galaxies by generating a repeated fluctuation in the potential well [2–5]. This result reconciles the dark matter density profile predicted in CDM that is steeply rising toward the center ("cuspy") [6–8] with observations which instead prefer a shallower density slope or even a constant dark matter density "core" [9–20]. Current simulations suggest that this process is the most effective in dwarf galaxies with stellar masses $\sim 10^8$ $M_\odot$ and halo virial masses of $\sim 10^{10}$ $M_\odot$. Below this mass, less star formation leads to less energy injection back to the interstellar gas in a galaxy, until there is simply not enough energy to alter the tightly bound cuspy dark matter profile. At higher masses, the deeper potential wells of galaxies like the Milky Way seem to prevent core formation [21, 22].

Early simulations that included only dark matter found that there should be many more satellites that orbit around our Milky Way galaxy within a CDM paradigm than we observe [23, 24]. Many of these satellites are expected to be "dark," unable to have formed stars due to photoevaporation of their gas when the Universe was reionized [25–29], though this process alone cannot bring the predicted number of massive, luminous satellites into an agreement with observation [30]. However, a simulation that includes baryons includes gas (by definition), which is able to cool itself (lose energy, primarily through radiation of photons). This is in contrast to dark matter, which is unable to cool. Cooling gas adds more mass to the center of the parent halo, creating stronger tidal forces that strip mass from satellite galaxies [31], and can also destroy satellites that pass too near the disk. Thus, the presence of a disk (which doesn't exist in a dark matter-only simulation) brings both the numbers and kinematics of satellite galaxies into agreement with observations [30, 32–35].

2 Should You Believe It?

The importance of baryons in creating realistic galaxies and overcoming the small-scale problems of CDM is now recognized by many simulators. Indeed, even simulators who do not have high enough resolution to resolve the processes that lead to dark matter core creation still find that inclusion of baryons can reconcile other outstanding challenges to CDM galaxy formation theory [36]. Thus, despite a range of star formation and feedback recipes, most simulators are now capable of simulating realistic galaxies that match a wide range of observed galaxy scaling relations (e.g., [37–39]).

A common question from the non-simulators at the Simons Symposium was, "What can be trusted in the simulations?" Much work has been done by simulators to address this same question. Two of the key things that go into cosmological galaxy simulations but that vary most widely from simulation to simulation are the efficiency at which stars are formed and the form of the energy feedback. A number of authors have now demonstrated that these two things are not independent; varying one will impact the other, with the net result being that galaxies converge to similar star formation rates and stellar masses because galaxies "self-regulate," i.e., a change in the star formation is counterbalanced by subsequent feedback and vice versa [40–46]. Figure 1 shows results from two different investigations of this topic. Self-regulation can occur as long as the resolution is high enough to capture the average densities in giant molecular clouds (GMCs), and, therefore, that the simulation is high enough resolution to have star formation limited to the scales of GMCs [47, 48]. Reference [49] recently demonstrated that self-regulation is limited to the regime of strong feedback (which most of the highest resolution simulations fall under), which regulates the gas supply available to turn into stars. It is because of galaxy self-regulation that most simulators operating at high resolutions generally find similar results and come to similar conclusions about galaxy evolution, despite varying parameters.

Another common question from non-simulators was, "What are the failings of the simulations?" This topic is always on the minds of simulators. In general, the biggest question right now is whether galaxy simulations can reproduce the range of diverse galaxy rotation curves that are currently observed (see Manoj Kaplinghat's summary in these proceedings). This question applies across a range of galaxy mass scales. At Milky Way masses, most simulations fail to create small stellar bulges, although the Milky Way and many of its largest spiral galaxy companions in the Local Volume seem to have small stellar bulges [50–52]. Simulations that do create small stellar bulges in Milky Way-mass galaxies don't seem to simultaneously be able to grow the disk as observed [53]. On the other hand, most high resolution Milky Way-mass simulations do not currently include supermassive black holes with AGN feedback. For a review on this topic, see Ref. [54].

In smaller galaxies, simulators seem to be able to create diffuse dwarfs, but multiple authors have noted that they have not created compact dwarfs [55, 56]. Possibly, this is due to small number statistics. Because zoomed simulations are computation-

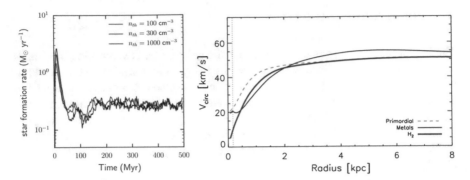

Fig. 1 *Left*: The SFR of a Milky Way-like simulated galaxy is unchanged as the density as which star formation is allowed to occur is changed (from 100 to 1000 amu cm^{-3}). From [44]. *Right*: A dwarf galaxy simulated using three different prescriptions for star formation and feedback yields a similar result in all cases. Shown here is the resulting rotation velocity. From [43]

ally expensive, the number of simulated galaxies is somewhat limited. However, it is also possible that we have entered a phase in which feedback is too strong, preventing simulations from forming the densest and thinnest galaxies we observe (e.g., [57]). The current inability to reproduce the full range of diverse galaxies is being actively addressed amongst the simulation community.

3 Implications for Non-CDM Models

Finally, a common misunderstanding was identified by Symposium participants: despite the fact that baryons within a CDM model can reconcile theory with many observations, *this does not mean that alternative dark matter models are not worth pursuing*. In fact, quite the opposite. A warm dark matter (WDM) model with baryons can still solve all of the small-scale problems and remain consistent with observations, as can a self-interacting dark matter (SIDM) model with baryons. There is no reason to believe that we understand all the properties of dark matter, and should therefore be pursuing a wide range of ideas. Thus, the question should really be: What are the predictions of alternative dark matter models with baryons included?

A sterile neutrino/dark fermion remains a viable dark matter candidate (see Kevork Abazajian's contribution in this proceedings). Some of the tightest constraints on WDM come from the abundance of low mass satellite galaxies [58–61] and the amount of small-scale structure in the Lyman-α forest [62–64]. In both cases, the WDM mass must be heavy enough that the data starts to look consistent with CDM, and the 3.5 keV line [65, 66] that is possibly produced by the decay of sterile neutrinos can still be made consistent with current observational constraints [67]. The allowed mass range of a WDM particle is thus very tight. Future x-ray telescopes should be able to resolve the 3.5 keV line, which will clarify its origin. Additional constraints on

WDM are likely to come from the earliest epoch of star formation [68–73]. Because structure formation is delayed in WDM models, a delay of star formation with respect to CDM expectations may point to WDM as the correct model.

SIDM, on the other hand, is a model for which the constraints have only been loosening over the past few years. After being initially invoked to solve the cusp-core problem [74], SIDM was quickly dismissed because it was believed to predict halo shapes that were more circular then observed [75, 76]. However, the question was revisited by Ref. [77], who demonstrated that a cross section for interaction, σ, of about $1\,cm^2/g$ (roughly the current limit in clusters, see Manoj Kaplinghat's contribution in this proceedings) does not lead to enough change in the halo shapes of clusters to significantly distinguish them from CDM. It has also been pointed out that the cross section for interaction is likely to be velocity dependent, with particles moving slower relative to each other more likely to interact. Reference [78] introduced such a model, allowing the constraints on cross section at dwarf scales to be revisited.

Reference [79] explored a dark matter-only SIDM simulation of a dwarf galaxy, at $9 \times 10^9\,M_\odot$ in halo mass. This halo was resimulated with SIDM cross sections of 0.1, 0.5, 1, 5, 10, and $50\,cm^2/g$. Figure 2 shows the resulting dark matter density profiles. In essence, they are all similar enough that it would be an observational challenge to try to distinguish between results from 0.5 to $50\,cm^2/g$. The $50\,cm^2/g$ model is currently the largest cross section explored to date. Even this large cross section cannot be ruled out, with its density profile being comparable to what is inferred in observed dwarf galaxies.

Reference [79] did not include baryonic physics, and the picture is altered further when baryons are considered. Reference [80] simulated a dwarf galaxy of comparable mass to the one run by [79], but with baryons. Reference [80] ran two models: an SIDM model with a cross section of $2\,cm^2/g$, and a standard CDM model. They discovered that the baryons begin to form a core before the SIDM model has enough time to start significantly scattering particles to create a core. Because of this, the resulting simulated dwarfs were identical (see Fig. 2 for their density profiles).

The largest cross section yet explored with baryons in this same galaxy mass range is $\sim 20\,cm^2/g$ (in the vdSIDMa simulation in Ref. [81]), and the results were entirely consistent with observations. Thus, there are currently no real constraints on the largest allowed cross section at dwarf galaxy scales. To constrain the particle physics models, two approaches should be taken: an observational approach and a simulation approach.

Observationally, there are a few hints that should be pursued further. First, in SIDM there are regimes where the baryons are likely to follow the dark matter distribution, e.g., in dark matter-dominated dwarf galaxies when the cross section is large [81]. The extent to which baryons trace DM needs to be explored in more detail, across a range of dwarf galaxy masses (from ultra-faints up to LMC-mass galaxies) and cross sections. Related, Ref. [82] found that SIDM satellites that fall into the Milky Way have their stellar orbits expanded as the halos get tidally stripped. Can the sizes of observed satellites be used to point to a DM model?

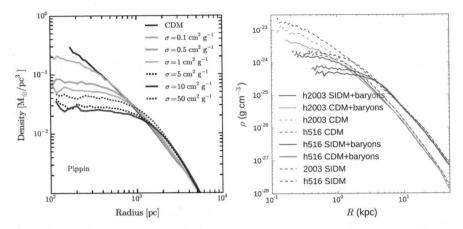

Fig. 2 *Left*: The density profiles for one dark matter halo of $9 \times 10^9\,M_\odot$ from Ref. [79]. The results are similar enough for $0.5 < \sigma < 50\,\mathrm{cm^2/g}$ that they would be observationally indistinguishable. *Right*: Simulation results from Ref. [80]. Two different dwarf galaxies are shown, run in varying models. The h516 dwarf galaxy has a similar mass to the one run by Ref. [79]. Because baryons are effective at creating dark matter cores at this mass, and because baryonic core creation occurs before many SIDM scatterings in this $2\,\mathrm{cm^2/g}$ model, both the SIDM and CDM models with baryons result in nearly identical density profiles. The h2003 dwarf, on the other hand, is low enough in stellar mass that baryonic effects don't create a strong core, while the SIDM model creates more of a dark matter core

Second, it has been noted that baryons do not provide enough energy to create cores in the ultra-faint dwarf galaxy range, with stellar masses $<10^5\,M_\odot$. However, SIDM could create cores in these small halos (see comparison of h2003 SIDM and CDM models in Fig. 2). Thus, measurements of the central densities and density slopes in the inner regions of dwarf galaxies are essential. Unfortunately, determining the central densities is a challenge. Even when scientists use the same data set, they have come to different conclusions about the presence of cores in dwarf galaxies (e.g., [83–85]) while ultra-faints contain many fewer stars and will thus be even more of a challenge. However, tackling the problem of how to determine central densities in dwarfs is an absolute priority in determining properties of dark matter–it could very well lead to a "smoking gun" that identifies or rules out a dark matter model.

From a simulation perspective, the ideal approach would be to crank up the SIDM interaction cross section and ask when galaxy formation breaks. That is, when do the simulation results stop being consistent with observations? A range of masses, from ultra-faint dwarf galaxies to the classical dwarf mass scale used above, should be investigated. However, resources are limited, both in terms of computing time and human resources. Individual simulations can require millions of CPU hours (and to fully explore SIDM, likely up to 100 million CPU hours would be needed). Meanwhile, few people are working on this topic given that the field is biased toward a preference for CDM with baryons.

4 Summary

In summary, there has been much progress in understanding the role of baryons in galaxy evolution in the past decade. To much of the astronomy community, this has solidified their confidence in the CDM model. However, it should primarily solidify their confidence in the ability of simulations to model baryonic physics. We still do not understand dark matter, and our favorite WIMP model continues to elude detection. Thus, we need to have an open mind about the possible properties of dark matter. A wide range of properties still waits to be explored in terms of consistency with galaxy evolution.

Acknowledgements Thank you to the Simons Foundation for hosting this Symposium, and to the organizers for bringing together a truly stimulating group of dark matter scientists. My work on baryons within a CDM model has been funded by NSF awards AST-1411399 and AST-1813871, and by the Space Telescope Science Institute awards HST-AR-13925 and HST-AR-14281.

References

1. R. Hlozek, J. Dunkley, G. Addison, J.W. Appel, J.R. Bond, C. Sofia Carvalho, S. Das, M.J. Devlin, R. Dünner, T. Essinger-Hileman, J.W. Fowler, P. Gallardo, A. Hajian, M. Halpern, M. Hasselfield, M. Hilton, A.D. Hincks, J.P. Hughes, K.D. Irwin, J. Klein, A. Kosowsky, T.A. Marriage, D. Marsden, F. Menanteau, K. Moodley, M.D. Niemack, M.R. Nolta, L.A. Page, L. Parker, B. Partridge, F. Rojas, N. Sehgal, B. Sherwin, J. Sievers, D.N. Spergel, S.T. Staggs, D.S. Swetz, E.R. Switzer, R. Thornton, E. Wollack, APJ **749**, 90 (2012). https://doi.org/10. 1088/0004-637X/749/1/90
2. J.I. Read, G. Gilmore, MNRAS **356**, 107 (2005). https://doi.org/10.1111/j.1365-2966.2004. 08424.x
3. R.S. de Souza, L.F.S. Rodrigues, E.E.O. Ishida, R. Opher, MNRAS **415**, 2969 (2011). https:// doi.org/10.1111/j.1365-2966.2011.18916.x
4. A. Pontzen, F. Governato, MNRAS **421**, 3464 (2012). https://doi.org/10.1111/j.1365-2966. 2012.20571.x
5. R. Teyssier, A. Pontzen, Y. Dubois, J.I. Read, MNRAS **429**, 3068 (2013). https://doi.org/10. 1093/mnras/sts563
6. J.F. Navarro, C.S. Frenk, S.D.M. White, APJ **490**, 493 (1997). https://doi.org/10.1086/304888
7. V. Springel, J. Wang, M. Vogelsberger, A. Ludlow, A. Jenkins, A. Helmi, J.F. Navarro, C.S. Frenk, S.D.M. White, MNRAS **391**, 1685 (2008). https://doi.org/10.1111/j.1365-2966.2008. 14066.x
8. J.F. Navarro, A. Ludlow, V. Springel, J. Wang, M. Vogelsberger, S.D.M. White, A. Jenkins, C.S. Frenk, A. Helmi, MNRAS **402**, 21 (2010). https://doi.org/10.1111/j.1365-2966.2009.15878.x
9. F.C. van den Bosch, B.E. Robertson, J.J. Dalcanton, W.J.G. de Blok, AJ **119**, 1579 (2000). https://doi.org/10.1086/301315
10. W.J.G. de Blok, S.S. McGaugh, V.C. Rubin, AJ **122**, 2396 (2001). https://doi.org/10.1086/ 323450
11. W.J.G. de Blok, A. Bosma, AAP **385**, 816 (2002). https://doi.org/10.1051/0004-6361: 20020080
12. J.D. Simon, A.D. Bolatto, A. Leroy, L. Blitz, APJ **596**, 957 (2003). https://doi.org/10.1086/ 378200
13. R.A. Swaters, B.F. Madore, F.C. van den Bosch, M. Balcells, APJ **583**, 732 (2003). https://doi. org/10.1086/345426

14. D.T.F. Weldrake, W.J.G. de Blok, F. Walter, MNRAS **340**, 12 (2003). https://doi.org/10.1046/j.1365-8711.2003.06170.x
15. R. Kuzio de Naray, S.S. McGaugh, W.J.G. de Blok, A. Bosma, APJS **165**, 461 (2006). https://doi.org/10.1086/505345
16. G. Gentile, P. Salucci, U. Klein, G.L. Granato, MNRAS **375**, 199 (2007). https://doi.org/10.1111/j.1365-2966.2006.11283.x
17. M. Spano, M. Marcelin, P. Amram, C. Carignan, B. Epinat, O. Hernandez, MNRAS **383**, 297 (2008). https://doi.org/10.1111/j.1365-2966.2007.12545.x
18. C. Trachternach, W.J.G. de Blok, F. Walter, E. Brinks, R.C. Kennicutt Jr., AJ **136**, 2720 (2008). https://doi.org/10.1088/0004-6256/136/6/2720
19. W.J.G. de Blok, F. Walter, E. Brinks, C. Trachternach, S. Oh, R.C. Kennicutt, AJ **136**, 2648 (2008). https://doi.org/10.1088/0004-6256/136/6/2648
20. S.H. Oh, C. Brook, F. Governato, E. Brinks, L. Mayer, W.J.G. de Blok, A. Brooks, F. Walter, AJ **142**, 24 (2011). https://doi.org/10.1088/0004-6256/142/1/24
21. A. Di Cintio, C.B. Brook, A.V. Macciò, G.S. Stinson, A. Knebe, A.A. Dutton, J. Wadsley, MNRAS **437**, 415 (2014). https://doi.org/10.1093/mnras/stt1891
22. T.K. Chan, D. Kereš, J. Oñorbe, P.F. Hopkins, A.L. Muratov, C.A. Faucher-Giguère, E. Quataert, MNRAS **454**, 2981 (2015). https://doi.org/10.1093/mnras/stv2165
23. B. Moore, S. Ghigna, F. Governato, G. Lake, T. Quinn, J. Stadel, P. Tozzi, APJL **524**, L19 (1999). https://doi.org/10.1086/312287
24. A. Klypin, A.V. Kravtsov, O. Valenzuela, F. Prada, APJ **522**, 82 (1999). https://doi.org/10.1086/307643
25. T. Quinn, N. Katz, G. Efstathiou, MNRAS **278**, L49 (1996)
26. A.A. Thoul, D.H. Weinberg, APJ **465**, 608 (1996). https://doi.org/10.1086/177446
27. R. Barkana, A. Loeb, APJ **523**, 54 (1999). https://doi.org/10.1086/307724
28. N.Y. Gnedin, APJ **542**, 535 (2000). https://doi.org/10.1086/317042
29. T. Okamoto, L. Gao, T. Theuns, MNRAS **390**, 920 (2008). https://doi.org/10.1111/j.1365-2966.2008.13830.x
30. A.M. Brooks, M. Kuhlen, A. Zolotov, D. Hooper, APJ **765**, 22 (2013). https://doi.org/10.1088/0004-637X/765/1/22
31. J. Peñarrubia, A.J. Benson, M.G. Walker, G. Gilmore, A.W. McConnachie, L. Mayer, MNRAS **406**, 1290 (2010). https://doi.org/10.1111/j.1365-2966.2010.16762.x
32. A. Zolotov, A.M. Brooks, B. Willman, F. Governato, A. Pontzen, C. Christensen, A. Dekel, T. Quinn, S. Shen, J. Wadsley, APJ **761**, 71 (2012)
33. A.M. Brooks, A. Zolotov, APJ **786** (2014)
34. A.R. Wetzel, P.F. Hopkins, J.h. Kim, C.A. Faucher-Giguère, D. Kereš, E. Quataert, APJL **827**, L23 (2016). https://doi.org/10.3847/2041-8205/827/2/L23
35. S. Garrison-Kimmel, A. Wetzel, J.S. Bullock, P.F. Hopkins, M. Boylan-Kolchin, C.A. Faucher-Giguère, D. Kereš, E. Quataert, R.E. Sanderson, A.S. Graus, T. Kelley, MNRAS **471**, 1709 (2017). https://doi.org/10.1093/mnras/stx1710
36. T. Sawala, C.S. Frenk, A. Fattahi, J.F. Navarro, R.G. Bower, R.A. Crain, C. Dalla Vecchia, M. Furlong, J.C. Helly, A. Jenkins, K.A. Oman, M. Schaller, J. Schaye, T. Theuns, J. Trayford, S.D.M. White, MNRAS **457**, 1931 (2016). https://doi.org/10.1093/mnras/stw145
37. C.B. Brook, G. Stinson, B.K. Gibson, J. Wadsley, T. Quinn, MNRAS **424**, 1275 (2012). https://doi.org/10.1111/j.1365-2966.2012.21306.x
38. M. Aumer, S.D.M. White, T. Naab, C. Scannapieco, MNRAS **434**, 3142 (2013). https://doi.org/10.1093/mnras/stt1230
39. M. Vogelsberger, S. Genel, D. Sijacki, P. Torrey, V. Springel, L. Hernquist, MNRAS **436**, 3031 (2013). https://doi.org/10.1093/mnras/stt1789
40. T.R. Saitoh, H. Daisaka, E. Kokubo, J. Makino, T. Okamoto, K. Tomisaka, K. Wada, N. Yoshida, PASJ **60**, 667 (2008)
41. P.F. Hopkins, E. Quataert, N. Murray, MNRAS **417**, 950 (2011). https://doi.org/10.1111/j.1365-2966.2011.19306.x

42. P.F. Hopkins, D. Narayanan, N. Murray, MNRAS **432**, 2647 (2013). https://doi.org/10.1093/mnras/stt723

43. C.R. Christensen, F. Governato, T. Quinn, A.M. Brooks, S. Shen, J. McCleary, D.B. Fisher, J. Wadsley, MNRAS **440**, 2843 (2014). https://doi.org/10.1093/mnras/stu399

44. S.M. Benincasa, J. Wadsley, H.M.P. Couchman, B.W. Keller, MNRAS **462**, 3053 (2016). https://doi.org/10.1093/mnras/stw1741

45. P.F. Hopkins, A. Wetzel, D. Keres, C.A. Faucher-Giguere, E. Quataert, M. Boylan-Kolchin, N. Murray, C.C. Hayward, S. Garrison-Kimmel, C. Hummels, R. Feldmann, P. Torrey, X. Ma, D. Angles-Alcazar, K.Y. Su, M. Orr, D. Schmitz, I. Escala, R. Sanderson, M.Y. Grudic, Z. Hafen, J.H. Kim, A. Fitts, J.S. Bullock, C. Wheeler, T.K. Chan, O.D. Elbert, D. Narananan, ArXiv e-prints (2017)

46. A. Pallottini, A. Ferrara, S. Bovino, L. Vallini, S. Gallerani, R. Maiolino, S. Salvadori, MNRAS **471**, 4128 (2017). https://doi.org/10.1093/mnras/stx1792

47. O. Agertz, A.V. Kravtsov, APJ **824**, 79 (2016). https://doi.org/10.3847/0004-637X/824/2/79

48. V.A. Semenov, A.V. Kravtsov, N.Y. Gnedin, APJ **826**, 200 (2016). https://doi.org/10.3847/0004-637X/826/2/200

49. V.A. Semenov, A.V. Kravtsov, N.Y. Gnedin, APJ **861**, 4 (2018). https://doi.org/10.3847/1538-4357/aac6eb

50. J. Kormendy, N. Drory, R. Bender, M.E. Cornell, APJ **723**(1), 54 (2010). https://doi.org/10.1088/0004-637X/723/1/54. http://adsabs.harvard.edu/abs/2010ApJ...723...54K

51. J. Shen, R.M. Rich, J. Kormendy, C.D. Howard, R. De Propris, A. Kunder, APJ **720**(1), L72 (2010). https://doi.org/10.1088/2041-8205/720/1/L72. http://adsabs.harvard.edu/abs/2010ApJ...720L.72S

52. D.B. Fisher, N. Drory, APJ **733**(2), L47 (2011). 10.1088/2041-8205/733/2/L47. http://arxiv.org/abs/1104.0020stacks.iop.org/2041-8205/733/i=2/a=L47?key=crossref.7e25a6051bda6f1c725b92db72faba5d

53. M. Aumer, S.D.M. White, T. Naab, MNRAS **441**(4), 3679 (2014). https://doi.org/10.1093/mnras/stu818. http://arxiv.org/abs/1404.6926mnras.oxfordjournals.org/cgi/doi/10.1093/mnras/stu818

54. A. Brooks, C. Christensen, in *Galactic Bulges, Astrophysics and Space Science Library*, vol. 418, ed. by E. Laurikainen, R. Peletier, D. Gadotti (Astrophysics and Space Science Library, 2016), p. 317. https://doi.org/10.1007/978-3-319-19378-6_12

55. I.M. Santos-Santos, A. Di Cintio, C.B. Brook, A. Macciò, A. Dutton, R. Domínguez-Tenreiro, MNRAS **473**, 4392 (2018). https://doi.org/10.1093/mnras/stx2660

56. S. Garrison-Kimmel, P.F. Hopkins, A. Wetzel, J.S. Bullock, M. Boylan-Kolchin, D. Keres, C.A. Faucher-Giguere, K. El-Badry, A. Lamberts, E. Quataert, R. Sanderson, ArXiv e-prints (2018)

57. K. El-Badry, A. Wetzel, M. Geha, P.F. Hopkins, D. Kereš, T.K. Chan, C.A. Faucher-Giguère, APJ **820**, 131 (2016). https://doi.org/10.3847/0004-637X/820/2/131

58. A.V. Macciò, F. Fontanot, MNRAS **404**, L16 (2010). https://doi.org/10.1111/j.1745-3933.2010.00825.x

59. E. Polisensky, M. Ricotti, PRD **83**(4), 043506 (2011). https://doi.org/10.1103/PhysRevD.83.043506

60. D. Anderhalden, J. Diemand, JCAP **4**, 009 (2013). https://doi.org/10.1088/1475-7516/2013/04/009

61. S. Horiuchi, P.J. Humphrey, J. Onorbe, K.N. Abazajian, M. Kaplinghat, S. Garrison-Kimmel, Phys. Rev. D **89**(2), 025017 (2014). https://doi.org/10.1103/PhysRevD.89.025017

62. M. Viel, J. Lesgourgues, M.G. Haehnelt, S. Matarrese, A. Riotto, Phys. Rev. Lett. **97**(7), 071301 (2006). https://doi.org/10.1103/PhysRevLett.97.071301

63. U. Seljak, A. Makarov, P. McDonald, H. Trac, Phys. Rev. Lett. **97**(19), 191303 (2006). https://doi.org/10.1103/PhysRevLett.97.191303

64. M. Viel, G.D. Becker, J.S. Bolton, M.G. Haehnelt, M. Rauch, W.L.W. Sargent, Phys. Rev. Lett. **100**(4), 041304 (2008). https://doi.org/10.1103/PhysRevLett.100.041304

65. A. Boyarsky, O. Ruchayskiy, D. Iakubovskyi, J. Franse, Phys. Rev. Lett. **113**, 251301 (2014). https://doi.org/10.1103/PhysRevLett.113.251301
66. E. Bulbul, M. Markevitch, A. Foster, R.K. Smith, M. Loewenstein, S.W. Randall, Astrophys. J. **789**, 13 (2014). https://doi.org/10.1088/0004-637X/789/1/13
67. K.N. Abazajian, Phys. Rev. Lett. **112**(16), 161303 (2014). https://doi.org/10.1103/PhysRevLett.112.161303
68. R. Barkana, Z. Haiman, J.P. Ostriker, APJ **558**, 482 (2001). https://doi.org/10.1086/322393
69. A. Mesinger, R. Perna, Z. Haiman, APJ **623**, 1 (2005). https://doi.org/10.1086/428770
70. R.S. de Souza, A. Mesinger, A. Ferrara, Z. Haiman, R. Perna, N. Yoshida, MNRAS **432**, 3218 (2013). https://doi.org/10.1093/mnras/stt674
71. F. Pacucci, A. Mesinger, Z. Haiman, MNRAS **435**, L53 (2013). https://doi.org/10.1093/mnrasl/slt093
72. F. Governato, D. Weisz, A. Pontzen, S. Loebman, D. Reed, A.M. Brooks, P. Behroozi, C. Christensen, P. Madau, L. Mayer, S. Shen, M. Walker, T. Quinn, B.W. Keller, J. Wadsley, MNRAS **448**, 792 (2015). https://doi.org/10.1093/mnras/stu2720
73. A. Chau, L. Mayer, F. Governato, APJ **845**, 17 (2017). https://doi.org/10.3847/1538-4357/aa7e74
74. D.N. Spergel, P.J. Steinhardt, Phys. Rev. Lett. **84**, 3760 (2000). https://doi.org/10.1103/PhysRevLett.84.3760
75. N. Yoshida, V. Springel, S.D.M. White, G. Tormen, APJL **544**, L87 (2000). https://doi.org/10.1086/317306
76. J. Miralda-Escudé, APJ **564**, 60 (2002). https://doi.org/10.1086/324138
77. A.H.G. Peter, M. Rocha, J.S. Bullock, M. Kaplinghat, Mon. Not. Roy. Astron. Soc. **430**, 105 (2013). https://doi.org/10.1093/mnras/sts535
78. A. Loeb, N. Weiner, Phys. Rev. Lett. **106**, 171302 (2011). https://doi.org/10.1103/PhysRevLett.106.171302
79. O.D. Elbert, J.S. Bullock, S. Garrison-Kimmel, M. Rocha, J. Oñorbe, A.H.G. Peter, Mon. Not. Roy. Astron. Soc. **453**(1), 29 (2015). https://doi.org/10.1093/mnras/stv1470
80. A.B. Fry, F. Governato, A. Pontzen, T. Quinn, M. Tremmel, L. Anderson, H. Menon, A.M. Brooks, J. Wadsley, MNRAS **452**, 1468 (2015). https://doi.org/10.1093/mnras/stv1330
81. M. Vogelsberger, J. Zavala, C. Simpson, A. Jenkins, MNRAS **444**, 3684 (2014). https://doi.org/10.1093/mnras/stu1713
82. G.A. Dooley, A.H.G. Peter, M. Vogelsberger, J. Zavala, A. Frebel, MNRAS **461**, 710 (2016). https://doi.org/10.1093/mnras/stw1309
83. M.A. Breddels, A. Helmi, R.C.E. van den Bosch, G. van de Ven, G. Battaglia, MNRAS **433**, 3173 (2013). https://doi.org/10.1093/mnras/stt956
84. M.G. Walker, J. Peñarrubia, APJ **742**, 20 (2011). https://doi.org/10.1088/0004-637X/742/1/20
85. L.E. Strigari, C.S. Frenk, S.D.M. White, ArXiv e-prints (2014)

Primordial Black Holes as Dark Matter and Generators of Cosmic Structure

Bernard Carr

Abstract Primordial black holes (PBHs) could provide the dark matter but a variety of constraints restrict the possible mass windows to $10^{16}-10^{17}$ g, $10^{20}-10^{24}$ g and $10-10^3 M_\odot$. The last possibility is of special interest in view of the recent detection of black hole mergers by LIGO. PBHs larger than $10^3 M_\odot$ might have important cosmological consequences even if they have only a small fraction of the dark matter density. In particular, they could generate cosmological structures either individually through the 'seed' effect or collectively through the 'Poisson' effect, thereby alleviating some problems associated with the standard cold dark matter scenario.

1 Introduction

Primordial black holes (PBHs) have been a source of interest for nearly 50 years [1], despite the fact that there is still no evidence for them. One reason for this interest is that only PBHs could be small enough for Hawking radiation to be important [2]. This has not yet been confirmed experimentally and there remain major conceptual puzzles associated with the process. Nevertheless, this discovery is generally recognised as one of the key developments in 20th century physics because it beautifully unifies general relativity, quantum mechanics and thermodynamics. The fact that Hawking was only led to this discovery by contemplating the properties of PBHs illustrates that it can be useful to study something even if it does not exist! But, of course, the situation is much more interesting if PBHs do exist.

PBHs smaller than about 10^{15} g would have evaporated by now with many interesting cosmological consequences [3]. Studies of such consequences have placed useful constraints on models of the early Universe and, more positively, evaporating PBHs have been invoked to explain certain features: for example, the extragalactic [4] and Galactic [5] γ-ray backgrounds, antimatter in cosmic rays [6], the annihilation line radiation from the Galactic centre [7], the reionisation of the pregalactic medium [8] and some short-period γ-ray bursts [9]. However, there are usually other possible

B. Carr (✉)
Queen Mary University of London, London, UK
e-mail: B.J.Carr@qmul.ac.uk

© Springer Nature Switzerland AG 2019
R. Essig et al. (eds.), *Illuminating Dark Matter*, Astrophysics and Space
Science Proceedings 56, https://doi.org/10.1007/978-3-030-31593-1_4

explanations for these features, so there is no definitive evidence for evaporating PBHs.

Attention has, therefore, shifted to the PBHs larger than 10^{15} g, which are unaffected by Hawking radiation. Such PBHs might have various astrophysical consequences, such as providing seeds for the supermassive black holes in galactic nuclei [10], the generation of large-scale structure through Poisson fluctuations [11] and important effects on the thermal and ionisation history of the Universe [12]. But perhaps the most exciting possibility is that they could provide the dark matter which comprises 25% of the critical density, an idea that goes back to the earliest days of PBH research [13]. Since PBHs formed in the radiation-dominated era, they are not subject to the well-known cosmological nucleosynthesis constraint that baryons can have at most 5% of the critical density [14]. They should, therefore, be classed as non-baryonic and behave like any other form of cold dark matter (CDM).

As with other CDM candidates, there is still no compelling evidence that PBHs provide the dark matter. There have been claims that the microlensing of quasars could indicate dark matter in jupiter-mass PBHs [15] but these are controversial. There was also a flurry of excitement about PBHs in 1997 when the MACHO microlensing results [16] suggested that the dark matter could be in compact objects of mass 0.5 $M_\odot$. Alternative dark matter candidates could be excluded and PBHs of this mass might naturally form at the quark-hadron phase transition at 10^{-5} s [17]. Subsequently, it was shown that such objects could comprise only 20% of the dark matter and indeed the entire mass range 10^{-7}–10 $M_\odot$ was excluded from providing the dark matter [18].

In recent decades attention has focused on other mass ranges in which PBHs could provide the dark matter and numerous constraints allow only three possibilities: the asteroid mass range (10^{16}–10^{17} g), the sublunar mass range (10^{20}–10^{24} g) and the intermediate mass black hole (IMBH) range (10–10^3 $M_\odot$). There is particular interest in the last possibility because the coalescing black holes detected by LIGO [19] could be of primordial origin, although this would not necessarily require the PBHs to provide all the dark matter. Also, PBHs could have important cosmological consequences even if they provide only a small fraction of the dark matter, so we explore this possibility below.

2 PBH Formation

PBHs could have been produced during the early Universe due to various mechanisms. Matching the cosmological density at a time t after the big bang with the density required to form a PBH of mass M implies that the PBH mass is comparable to the horizon mass at formation [20, 21]:

$$M \sim \frac{c^3 t}{G} \sim 10^{15} \left(\frac{t}{10^{-23} \text{s}} \right) \text{g}. \tag{1}$$

Hence PBHs could span an enormous mass range: those formed at the Planck time (10^{-43} s) would have the Planck mass (10^{-5} g), whereas those formed at 1 s would be as large as $10^5 \, M_\odot$. By contrast, black holes forming at the present epoch (e.g. in the final stages of stellar evolution) could never be smaller than about 1 $M_\odot$. In some circumstances PBHs may form over an extended period, corresponding to a wide range of masses, but their spectrum could be extended even if they form at a single epoch.

As discussed in numerous papers, the quantum fluctuations arising in various inflationary scenarios are a possible source of PBHs. In some of these scenarios, the fluctuations generated by inflation are 'blue' (i.e. decrease with increasing scale) and this means that the PBHs form shortly after reheating. Others involve some form of 'designer' inflation, in which the power spectrum of the fluctuations — and hence PBH production — peaks on some scale. In other scenarios, the fluctuations have a 'running index', so that the amplitude increases on smaller scales but not according to a simple power law. PBH formation may also occur due to some sort of parametric resonance effect before reheating, in which case, the fluctuations tend to peak on a scale associated with reheating. This is usually very small but several scenarios involve a secondary inflationary phase which boosts this scale into the macroscopic domain. Detailed references for all these models can be found in Ref. [3].

Whatever the source of the inhomogeneities, PBH formation would be enhanced if there was a reduction in the pressure at some epoch - for example, at the QCD era [22] or if the early Universe went through a dust-like phase as a result of being dominated by nonrelativistic particles for a period [23, 24] or undergoing slow reheating after inflation [25, 26]. Another possibility is that PBHs might have formed spontaneously at some sort of phase transition, even if there were no prior inhomogeneities, for example from the collisions of bubbles of broken symmetry or the collapse of cosmic strings or domain walls. References for such models can be found in Ref. [3].

The fraction of the mass of the Universe in PBHs is time-dependent but its value at the PBH formation epoch is of particular interest. If the PBHs formed at a redshift z or time t and contribute a fraction $f(M)$ of the dark matter on some mass-scale M, then the collapse fraction is [27]

$$\beta(M) = f(M) \left(\frac{1+z}{1+z_{eq}} \right) \sim 10^{-6} f(M) \left(\frac{t}{1 \, \text{s}} \right)^{1/2} \sim 10^{-18} f(M) \left(\frac{M}{10^{15} \text{g}} \right)^{1/2},$$
(2)

where we assume the PBHs form in the radiation-dominated era, $z_{eq} \approx 4000$ is the redshift of matter-radiation equality, and we use Eq. (1) at the last step. The $(1 + z)$ factor arises because the radiation density scales as $(1 + z)^4$, whereas the PBH density scales as $(1 + z)^3$. Any limit on $f(M)$ (e.g. $f \leq 1$ for $M > 10^{15}$ g), therefore, places a constraint on $\beta(M)$, which is necessarily tiny.

On the other hand, one also *expects* the collapse fraction to be small. For example, if the PBHs form from primordial inhomogeneities which are Gaussian with rms amplitude $\delta_H(M)$ at the horizon epoch, one predicts [27]

$$\beta(M) \approx \mathrm{erfc}\left(\frac{\delta_c}{\sqrt{2}\delta_H(M)}\right),\tag{3}$$

where 'erfc' is the complimentary error function and $\delta_c \approx 0.4$ is the threshold for collapse against the pressure [28, 29]. In a dust era, the collapse fraction is $\beta \sim 0.02\delta_H(M)^5$, corresponding to the probability of sufficient spherical symmetry, but this is still small [30, 31]. In the other scenarios, β depends upon some cosmological parameter (eg. the string tension or bubble formation rate).

3 Constraints on Nonevaporated Black Holes

The constraints on $f(M)$, the fraction of the halo in PBHs of mass M, are summarised in Fig. 1, which is taken from Ref. [32]. All the limits assume that the PBHs cluster in the Galactic halo in the same way as other forms of CDM. The effects are extragalactic γ-rays from evaporations (EG) [3], femtolensing of γ-ray bursts (F) [33], white dwarf explosions (WD) [34], neutron star captures (NS) [35], Kepler microlensing of stars (K) [36], MACHO/EROS/OGLE microlensing of stars (ML) [18] and quasar microlensing (broken line) (ML) [37], survival of a star cluster in Eridanus II (E) [38], wide binary disruption (WB) [39], dynamical friction on halo objects (DF) [40], millilensing of quasars (mLQ) [41], generation of large-scale structure through Poisson fluctuations (LSS) [11], and accretion effects (WMAP, FIRAS) [12].

As indicated by the arrows in Fig. 1, the permitted mass windows for $f \sim 1$ are: (A) the intermediate mass range ($10–10^3\,M_\odot$; (B) the sublunar mass range ($10^{20}–10^{26}$ g); and (C) the asteroid mass range ($10^{16}–10^{17}$ g). However, there are further limits since Fig. 1 was produced and some people claim that even these windows are excluded. For example, scenario C may be ruled out by Galactic γ-ray observations [42]. One problem with scenario A is that such objects would disrupt wide binaries in the Galactic disc. It was originally claimed that this would exclude objects above $400\,M_\odot$ [39] but more recent studies may reduce this mass [43], so the narrow window between the microlensing and wide binary bounds is shrinking. There are new microlensing constraints in the lunar mass range from the Subaru telescope [44]. Also, the CMB accretion constraints have been revised and are now weaker [45], although there are new accretion limits from X-ray observations [46, 47]. Two talks at this symposium report interesting new constraints associated with tidal streams [48] and lensing substructure [49].

The PBHs in either scenario A and B could be generated by inflation but theorists are split as to which window they favour. For example, Inomata et al. [50] argue that double inflation can produce a peak at around 10^{20} g, while Clesse and Garcia-Bellido [51] argue that hybrid inflation can produce a peak at around $10M_\odot$, this being favoured by the LIGO results. On the other hand, a peak at around this mass could also be produced by a reduction in the pressure at the quark-hadron phase transition [52]. In this sense, there is a parallel with the search for particle dark matter candidates, where there is a split between groups searching for light and heavy candidates.

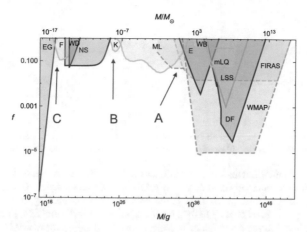

Fig. 1 Constraints on $f(M)$ from Ref. [32] for a variety of evaporation (magenta), dynamical (red), lensing (cyan), large-scale structure (green) and accretion (orange) effects. Only the strongest constraint is usually included in each mass range; the accretion limits are shown with broken lines since they are highly model-dependent. The arrows indicate the three mass windows where f can be close to 1

The constraints discussed above assume that the PBH mass function is monochromatic (i.e. with a width $\Delta M \sim M$). However, there are many scenarios in which one would expect the mass function to be extended. For example. inflation tends to produce a lognormal mass function [53] and critical collapse generates an extended low mass tail [54]. In the context of the dark matter problem, this is a two-edged sword [32]. On the one hand, it means that the *total* PBH density may suffice to explain the dark matter, even if the density in any particular mass band is small and within the observational bounds discussed above. On the other hand, even if PBHs can provide all the dark matter at some mass-scale, the extended mass function may still violate the constraints at some other scale. This issue has been addressed in a number of recent papers [55, 56], though with somewhat different conclusions.

4 Effects of PBHs on Cosmic Structures

PBHs of mass m provide a source of fluctuations for objects of mass M in two ways: (1) via the seed effect, in which the Coulomb effect of a *single* black hole generates an initial density fluctuation m/M; (2) via the Poisson effect, in which the $\sqrt{N}$ fluctuation in the number of black holes generates an initial density fluctuation $(fm/M)^{1/2}$. Both types of fluctuations then grow through gravitational instability to bind regions of mass M. Each of these proposals has a long history and detailed references can be found in Ref. [57]. The relationship between the two mechanisms is subtle, so we will consider both of them below and determine the dominant one for each mass-scale.

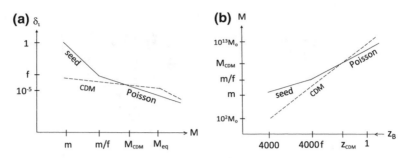

Fig. 2 a Form of initial fluctuation δ_i as a function of M for the seed and Poisson effect with fixed f, the first dominating at small M if f is small but the second always dominating if $f \sim 1$. **b** Mass M binding at redshift z_B for fixed f, the Poisson effect dominating for low z if f is small but at all z if $f \sim 1$. Also shown by dashed lines are the forms for δ_i and $M(z_B)$ predicted by the CDM model, this indicating the range $M > M_{CDM}$ and $z_B < z_{CDM}$ for which CDM fluctuations dominate. From Ref. [57]

If the PBHs have a single mass m, the initial fluctuation in the matter density on a scale M is

$$\delta_i \approx m/M \text{ (seed)}, \quad (fm/M)^{1/2} \text{ (Poisson)}, \qquad (4)$$

where M excludes the radiation content. If PBHs provide the dark matter, $f \sim 1$ and the Poisson effect dominates for all M but we also consider scenarios with $f \ll 1$. The Poisson effect then dominates for $M > m/f$ and the seed effect for $M < m/f$. Indeed, the first expression in (4) only applies for $f < m/M$, since otherwise a region of mass M would be expected to contain more than one black hole. The dependence of δ_i on M is indicated in Fig. 2a. The fluctuation grows as $(1 + z)^{-1}$ from the redshift of matter-radiation equality, $z_{eq} \approx 4000$, until it binds when $\delta \approx 1$. Therefore the mass binding at redshift z_B is

$$M \approx 4000\, m z_B^{-1} \text{ (seed)}, \quad 10^7 f m z_B^{-2} \text{ (Poisson)}, \qquad (5)$$

as illustrated in Fig. 2b. The CDM fluctuations are shown for comparison. These always dominate at sufficiently large scales but the PBHs provide an extra peak in the power spectrum on small scales.

One can place interesting upper limits on $f(m)$ by requiring that various types of structure do not form too early. But one can also take a more positive approach, exploring the possibility that PBHs may have *helped* the formation of these objects, thereby complementing the standard CDM scenario of structure formation. If the PBHs have a monochromatic mass function and provide *all* the dark matter ($f \sim 1$), then the Poisson effect dominates on all scales and various astrophysical constraints discussed above require $m < 10^3 M_\odot$. This implies that PBHs can only bind sub-galactic masses but still allows them to play a role in producing the first bound baryonic clouds or the SMBHs which power quasars. For $f \ll 1$, the seed effect dominates on small scales and can bind a region of up to 4000 times the PBH mass. Most galaxies contain central supermassive black holes with a mass proportional to the bulge mass and this correlation is naturally explained by the seed effect if the black holes are primordial with an extended mass function. However, limits on the μ-distortion in the CMB due to the dissipation of fluctuations before decoupling exclude PBHs larger than $10^5 M_\odot$ unless one invokes non-Gaussian fluctuations or accretion [58].

5 LIGO Gravitational Wave Limits

The proposal that the dark matter could comprise PBHs in the intermediate mass range has attracted much attention recently as a result of the LIGO detections of merging binary black holes with mass around $30\,M_\odot$ [59–61]. Since the black holes are larger than initially expected, it has been suggested that they could represent a new population. One possibility is that they were of Population III origin (i.e. forming between decoupling and galaxies). The suggestion that LIGO might detect gravitational waves from coalescing intermediate mass Population III black holes was first made more than 30 years ago [62] and—rather remarkably—Kinugawa et al. predicted a Population III coalescence peak at $30 M_\odot$ shortly before the first LIGO detection [63].

Another possibility – more relevant to the present considerations – is that the LIGO black holes were primordial, as first discussed in Ref. [64]. This does not necessarily require the PBHs to provide *all* the dark matter. While several authors have made this connection [65, 66], the predicted merger rate depends on many uncertain astrophysical factors and others argue that the PBH density could be much less than the dark matter density [67, 68]. Note that the PBH density should peak at a lower mass than the coalescence signal for an extended PBH mass function, since the amplitude of the gravitational waves scales as the black hole mass. Indeed, Clesse and Garcia-Bellido argue that a lognormal distribution centred at around $3 M_\odot$ would naturally explains both the dark matter and the LIGO bursts without violating any of the current PBH constraints [66].

A population of massive PBHs would also be expected to generate a stochastic background of gravitational waves [69], whether or not they form binaries. If the PBHs have an extended mass function, incorporating both dark matter at the low end and galactic seeds at the high end, this would have important implications for the predicted gravitational wave background. Theorists usually focus on the gravitational waves generated by either stellar black holes (detectable by LIGO) or supermassive black holes (detectable by LISA). However, with an extended PBH mass function, the gravitational wave background should encompass both these limits and also every intermediate frequency.

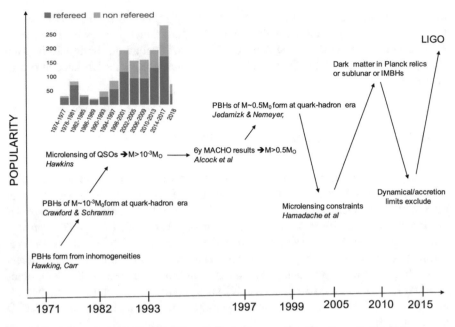

Fig. 3 Evolution of popularity of PBHs as indicated by annual publication rate (top left)

6 Summary

In recent years PBHs have been invoked for three purposes: (1) to explain the dark matter; (2) to provide a source of LIGO coalescences; (3) to alleviate some of the problems associated with the CDM scenario. In principle, these are distinct roles and any one of them would justify the study of PBHs. On the other hand, if PBHs have an extended mass function, they could play all three roles.

As regards (1), there are only a few mass ranges in which PBHs could provide the dark matter. We have focused particularly on the intermediate mass range $10\,M_\odot < M < 10^3\,M_\odot$, because this may be relevant to (2), but the sublunar range $10^{20}-10^{24}$ g also remains viable. The asteroid range $10^{16}-10^{17}$ g is probably the least plausible. We have not discussed the possibility that stable Planck mass relics of PBH evaporations provide the dark matter [70]. This scenario cannot be excluded but it is impossible to test since Planck scale relics would be undetectable except through their gravitational effects.

Presumably, most participants at this meeting would prefer the dark matter to be elementary particles rather than PBHs, so it may be reassuring that for most of the last 50 years the study of PBHs has been a minority interest. On the other hand, as illustrated in Fig. 3, PBHs have become increasingly popular in recent years, at least as measured by the annual publication rate on the topic. Indeed, turning to role (3), perhaps the most important point to emphasise, is that PBHs in the intermediate to supermassive mass range could play an important cosmological role even if they do *not* provide the dark matter. Perhaps this also applies for the particle candidates. Few people would now argue that neutrinos provide the dark matter but they are still extraordinarily important.

Acknowledgements This talk is dedicated to the memory of my friend and mentor Stephen Hawking. If PBHs turn out to exist, then his pioneering work on this topic will have been one of his most prescient and important scientific contributions. I thank the Simons Foundation for their generous hospitality at this conference and my many PBH coathors over 45 years for an enjoyable collaboration.

References

1. I. YaB Zel'dovich, Novikov, Sov. Astron. **10**, 602 (1967)
2. S.W. Hawking, Nature **248**, 30 (1974). https://doi.org/10.1038/248030a0
3. B.J. Carr, K. Kohri, Y. Sendouda, J. Yokoyama, Phys. Rev. D **81**, 104019 (2010). https://doi.org/10.1103/PhysRevD.81.104019
4. D.N. Page, S.W. Hawking, Astrophys. J. **206**, 1 (1976). https://doi.org/10.1086/154350
5. R. Lehoucq, M. Casse, J.M. Casandjian, I. Grenier, Astron. Astrophys. **502**, 37 (2009). https://doi.org/10.1051/0004-6361/200911961
6. A. Barrau, Astropart. Phys. **12**, 269 (2000). https://doi.org/10.1016/S0927-6505(99)00103-6
7. C. Bambi, A.D. Dolgov, A.A. Petrov, Phys. Lett. B **670**, 174 (2008). https://doi.org/10.1016/j.physletb.2009.10.053, https://doi.org/10.1016/j.physletb.2008.10.057. [Erratum: Phys. Lett. B681,504(2009)]
8. K.M. Belotsky, A.A. Kirillov, JCAP **1501**(01), 041 (2015). https://doi.org/10.1088/1475-7516/2015/01/041
9. D.B. Cline, D.A. Sanders, W. Hong, Astrophys. J. **486**, 169 (1997). https://doi.org/10.1086/304480
10. R. Bean, J. Magueijo, Phys. Rev. D **66**, 063505 (2002). https://doi.org/10.1103/PhysRevD.66.063505
11. N. Afshordi, P. McDonald, D.N. Spergel, Astrophys. J. Lett. **594**, L71 (2003). https://doi.org/10.1086/378763
12. M. Ricotti, J.P. Ostriker, K.J. Mack, Astrophys. J. **680**, 829 (2008). https://doi.org/10.1086/587831
13. G.F. Chapline, Nature **253**, 251 (1975). https://doi.org/10.1038/253251a0
14. R.H. Cyburt, B.D. Fields, K.A. Olive, Phys. Lett. B **567**, 227 (2003). https://doi.org/10.1016/j.physletb.2003.06.026
15. M.R.S. Hawkins, Nature **366**, 242 (1993). https://doi.org/10.1038/366242a0
16. C. Alcock et al., Astrophys. J. **486**, 697 (1997). https://doi.org/10.1086/304535
17. K. Jedamzik, Phys. Rep. **307**, 155 (1998). https://doi.org/10.1016/S0370-1573(98)00067-2
18. P. Tisserand et al., Astron. Astrophys. **469**, 387 (2007). https://doi.org/10.1051/0004-6361:20066017
19. B.P. Abbott et al., Phys. Rev. Lett. **116**(6), 061102 (2016). https://doi.org/10.1103/PhysRevLett.116.061102
20. S. Hawking, Mon. Not. Roy. Astron. Soc. **152**, 75 (1971)
21. B.J. Carr, S.W. Hawking, Mon. Not. Roy. Astron. Soc. **168**, 399 (1974)
22. M. Crawford, D.N. Schramm, Nature **298**, 538 (1982). https://doi.org/10.1038/298538a0
23. A.G. Polnarev, M.Y. Khlopov, Sov. Astron. **26**, 391 (1982)
24. B. Carr, T. Tenkanen, V. Vaskonen, Phys. Rev. D **96**(6), 063507 (2017). https://doi.org/10.1103/PhysRevD.96.063507
25. M.Y. Khlopov, B.A. Malomed, Y.B. Zeldovich, Mon. Not. Roy. Astron. Soc. **215**, 575 (1985)

26. B. Carr, K. Dimopoulos, C. Owen, T. Tenkanen, Phys. Rev. D **97**(12), 123535 (2018). https://doi.org/10.1103/PhysRevD.97.123535
27. B.J. Carr, Astrophys. J. **201**, 1 (1975). https://doi.org/10.1086/153853
28. I. Musco, J.C. Miller, L. Rezzolla, Class. Quant. Grav. **22**, 1405 (2005). https://doi.org/10.1088/0264-9381/22/7/013
29. T. Harada, C.M. Yoo, K. Kohri, Phys. Rev. D **88**(8), 084051 (2013). https://doi.org/10.1103/PhysRevD.88.084051, https://doi.org/10.1103/PhysRevD.89.029903. [Erratum: Phys. Rev. D **89**, no.2,029903(2014)]
30. M.Y. Khlopov, A.G. Polnarev, Phys. Lett. B **97**, 383 (1980). https://doi.org/10.1016/0370-2693(80)90624-3
31. T. Harada, C.M. Yoo, K. Kohri, K.I. Nakao, Phys. Rev. D **96**(8), 083517 (2017). https://doi.org/10.1103/PhysRevD.96.083517
32. B. Carr, F. Kuhnel, M. Sandstad, Phys. Rev. D **94**(8), 083504 (2016). https://doi.org/10.1103/PhysRevD.94.083504
33. A. Barnacka, J.F. Glicenstein, R. Moderski, Phys. Rev. D **86**, 043001 (2012). https://doi.org/10.1103/PhysRevD.86.043001
34. P.W. Graham, S. Rajendran, J. Varela, Phys. Rev. D **92**(6), 063007 (2015). https://doi.org/10.1103/PhysRevD.92.063007
35. F. Capela, M. Pshirkov, P. Tinyakov, Phys. Rev. D **87**(12), 123524 (2013). https://doi.org/10.1103/PhysRevD.87.123524
36. K. Griest, A.M. Cieplak, M.J. Lehner, Astrophys. J. **786**(2), 158 (2014). https://doi.org/10.1088/0004-637X/786/2/158
37. E. Mediavilla, J.A. Munoz, E. Falco, V. Motta, E. Guerras, H. Canovas, C. Jean, A. Oscoz, A.M. Mosquera, Astrophys. J. **706**, 1451 (2009). https://doi.org/10.1088/0004-637X/706/2/1451
38. T.D. Brandt, Astrophys. J. **824**(2), L31 (2016). https://doi.org/10.3847/2041-8205/824/2/L31
39. D.P. Quinn et al., Mon. Not. Roy. Astron. Soc. Lett. **396**, L11 (2009). https://doi.org/10.1111/j.1745-3933.2009.00652.x
40. B.J. Carr, M. Sakellariadou, Astrophys. J. **516**, 195 (1999). https://doi.org/10.1086/307071
41. P.N. Wilkinson et al., Phys. Rev. Lett. **86**, 584 (2001). https://doi.org/10.1103/PhysRevLett.86.584
42. B.J. Carr, K. Kohri, Y. Sendouda, J. Yokoyama, Phys. Rev. D **94**(4), 044029 (2016)
43. M.A. Monroy-Rodríguez, C. Allen, Astrophys. J. **790**(2), 159 (2014). https://doi.org/10.1088/0004-637X/790/2/159
44. H. Niikura, M. Takada, N. Yasuda, R.H. Lupton, T. Sumi, S. More, A. More, M. Oguri, M. Chiba, Nat. Astron. **3**, 524 (2019)
45. Y. Ali-Haïmoud, M. Kamionkowski, Phys. Rev. D **95**(4), 043534 (2017). https://doi.org/10.1103/PhysRevD.95.043534
46. V. Poulin, P.D. Serpico, F. Calore, S. Clesse, K. Kohri, Phys. Rev. D **96**(8), 083524 (2017). https://doi.org/10.1103/PhysRevD.96.083524
47. Y. Inoue, A. Kusenko, JCAP **1710**(10), 034 (2017). https://doi.org/10.1088/1475-7516/2017/10/034
48. J. Bovy, D. Erkal, J.L. Sanders, Mon. Not. Roy. Astron. Soc. **466**(1), 628 (2017). https://doi.org/10.1093/mnras/stw3067
49. Y.D. Hezaveh et al., Astrophys. J. **823**(1), 37 (2016). https://doi.org/10.3847/0004-637X/823/1/37
50. K. Inomata, M. Kawasaki, K. Mukaida, Y. Tada, T.T. Yanagida, Phys. Rev. D **96**(4), 043504 (2017). https://doi.org/10.1103/PhysRevD.96.043504
51. S. Clesse, J. García-Bellido, Phys. Rev. D **92**(2), 023524 (2015). https://doi.org/10.1103/PhysRevD.92.023524
52. C.T. Byrnes, M. Hindmarsh, S. Young, M.R.S. Hawkins, JCAP **1808**, 041 (2018)
53. A. Dolgov, J. Silk, Phys. Rev. D **47**, 4244 (1993). https://doi.org/10.1103/PhysRevD.47.4244
54. J. Yokoyama, Phys. Rev. D **58**, 107502 (1998). https://doi.org/10.1103/PhysRevD.58.107502
55. A.M. Green, Phys. Rev. D **94**(6), 063530 (2016). https://doi.org/10.1103/PhysRevD.94.063530

56. B. Carr, M. Raidal, T. Tenkanen, V. Vaskonen, H. Veerme, Phys. Rev. D **96**(2), 023514 (2017). https://doi.org/10.1103/PhysRevD.96.023514
57. B. Carr, J. Silk (2018). https://doi.org/10.1093/mnras/sty1204
58. T. Nakama, B. Carr, J. Silk, Phys. Rev. D **97**(4), 043525 (2018). https://doi.org/10.1103/PhysRevD.97.043525
59. B.P. Abbott et al., Phys. Rev. X **6**(4), 041015 (2016). https://doi.org/10.1103/PhysRevX.6.041015
60. B.P. Abbott et al., Phys. Rev. Lett. **116**(24), 241102 (2016). https://doi.org/10.1103/PhysRevLett.116.241102
61. B.P. Abbott et al., Phys. Rev. Lett. **116**(24), 241103 (2016). https://doi.org/10.1103/PhysRevLett.116.241103
62. J.R. Bond, B.J. Carr, Mon. Not. Roy. Astron. Soc. **207**, 585 (1984). https://doi.org/10.1093/mnras/207.3.585
63. T. Kinugawa, K. Inayoshi, K. Hotokezaka, D. Nakauchi, T. Nakamura, Mon. Not. Roy. Astron. Soc. **442**(4), 2963 (2014). https://doi.org/10.1093/mnras/stu1022
64. T. Nakamura, M. Sasaki, T. Tanaka, K.S. Thorne, Astrophys. J. Lett. **487**, L139 (1997). https://doi.org/10.1086/310886
65. S. Bird, I. Cholis, J.B. Muñoz, Y. Ali-Haïmoud, M. Kamionkowski, E.D. Kovetz, A. Raccanelli, A.G. Riess, Phys. Rev. Lett. **116**(20), 201301 (2016). https://doi.org/10.1103/PhysRevLett.116.201301
66. S. Clesse, J. García-Bellido, Phys. Dark Univ. **15**, 142 (2017). https://doi.org/10.1016/j.dark.2016.10.002
67. M. Sasaki, T. Suyama, T. Tanaka, S. Yokoyama, Phys. Rev. Lett. **117**(6), 061101 (2016). https://doi.org/10.1103/PhysRevLett.121.059901, https://doi.org/10.1103/PhysRevLett.117.061101. [Erratum: Phys. Rev. Lett.121, no.5,059901(2018)]
68. T. Nakamura, et al., PTEP **2016**(9), 093E01 (2016). https://doi.org/10.1093/ptep/ptw127
69. B.J. Carr, Astron. Astrophys. **89**, 6 (1980)
70. J.H. MacGibbon, Nature **329**, 308 (1987). https://doi.org/10.1038/329308a0

Quantum Metrology Techniques
for Axion Dark Matter Detection

Aaron S. Chou

Abstract Quantum metrology techniques can be used to dramatically improve the signal-to-noise ratio in experimental searches for dark matter axion waves. We first briefly review the cavity haloscope technique including quantum-limited measurements in the Glauber coherent state basis. Quantum non-demolition measurements in the Fock basis are then shown to offer much reduced background rates. Finally, we show that by preparing the cavity photon mode in a Fock state, the axion signal can be enhanced by stimulated emission.

Keywords Axion · Quantum · Qubits · Non-demolition · Fock · Coherent

1 Review of Haloscopes Using Quantum-Limited Amplifiers

For particulate dark matter masses below around 100 eV, the occupation number per mode volume $\lambda_{\text{deBroglie}}^3$ exceeds unity, and dark matter must then be bosonic. For the much lower masses expected for QCD axions, the mode occupation numbers become macroscopic and the local axion dark matter may be described as a stationary classical sine wave

$$\theta = \theta_0 e^{im_a t} \tag{1}$$

with amplitude

$$\theta_0 = \sqrt{\frac{2\rho_a}{\Lambda_{\text{QCD}}^4}} \approx 3.7 \times 10^{-19} \, \text{rad} \tag{2}$$

determined by how far the local axion dark matter density $\rho_a \approx 300 \, \text{MeV/cm}^3$ enables the field to coherently climb up the Peccei–Quinn tilted potential well char-

A. S. Chou (✉)
Fermilab, Batavia, IL, USA
e-mail: achou@fnal.gov
URL: http://home.fnal.gov/~achou

© Springer Nature Switzerland AG 2019
R. Essig et al. (eds.), *Illuminating Dark Matter*, Astrophysics and Space
Science Proceedings 56, https://doi.org/10.1007/978-3-030-31593-1_5

41

acterized by potential energy density $\Lambda_{\text{QCD}}^4 \approx (77\,\text{MeV})^4$. For Doppler broadening by the virial velocity and velocity spread $v \approx \Delta v \approx 10^{-3}c$, this dark matter sine wave has a momentum spread $\Delta p = 10^{-3}m_a$ and is coherent over spatial scales $1/\Delta p = 10^3/m_a$. The kinetic broadening about the dominant rest mass gives a signal linewidth $\Delta f = 10^{-6}m_a/2\pi$, corresponding to a quality factor $Q_a \approx 10^6$.

In the presence of a laboratory magnetic field B_0, the bilinear interaction Lagrangian

$$\mathcal{L}_{\text{int}} = g\theta \mathbf{E} \cdot \mathbf{B}_0 \tag{3}$$

enables the axion sine wave θ to drive a resonant mode of an electromagnetic cavity via its electric field $\mathbf{E}$. The power delivered to the cavity is

$$P_s = \int dV \mathbf{J}_a(x) \cdot \mathbf{E}(x) \tag{4}$$

where the space-filling exotic current density is

$$\mathbf{J}_a(x) = ig\theta \mathbf{B}_0(x)m_a \tag{5}$$

and $\mathbf{E}(x)$ is the electric field (including its spatial profile) of the cavity mode being driven. Generally one designs an experiment in which the lowest order cavity mode with uniformly oriented $\mathbf{E}$ in each half cycle is used in conjunction with a spatially constant B_0 field in order to generate the maximum overlap integral so that different parts of the current do not push the mode in opposite directions.

Through the Maxwell source equation

$$\nabla \times \mathbf{H}_c - \partial_t \mathbf{D}_c = J_a, \tag{6}$$

the electric displacement field of the signal builds up linearly in the cavity over Q_c oscillations to a peak value

$$|\mathbf{D}_c| = 2g\theta_0 B_0 Q_c \tag{7}$$

as can be seen by comparing the energy loss rate to the energy delivery rate

$$\langle \frac{1}{2}\mathbf{D}_c \cdot \mathbf{E}_c \rangle \Gamma = \langle \mathbf{J}_a \cdot \mathbf{E}_c \rangle \tag{8}$$

for decay rate $\Gamma = \omega/Q_c = m_a/Q_c$. This electric field corresponds to an average stored energy

$$U_c = \frac{1}{2}\langle \mathbf{D}_c \cdot \mathbf{E}_c \rangle V = \frac{1}{\epsilon}(g\theta_0)^2 B_0^2 V Q_c^2. \tag{9}$$

where ϵ is the electric permittivity of the bulk material of the cavity resonator, and a cycle averaging factor of $\langle \sin^2(m_a t) \rangle = 1/2$ has been included. This energy is lost on a time scale equal to the cavity lifetime and so the power delivered to the cavity into all loss channels (including the power possibly siphoned to a readout amplifier) is

$$P_s = U_c m_a / Q_c = \frac{1}{\epsilon}(g\theta_0)^2 B_0^2 V m_a Q_c \tag{10}$$

which is Sikivie's formula [1]. The signal photon rate is then

$$R_s = P_s / m_a = \frac{1}{\epsilon}(g\theta_0)^2 B_0^2 V Q_c. \tag{11}$$

If the signal is read out with a phase-preserving amplifier which obeys the standard quantum limit (SQL), then the effective noise photon rate (whose shot noise equals the readout noise variance) is simply one photon per resolved mode with equal contributions from the zero-point noise of the cavity and of the amplifier [2]. One may view this noise as the variance of the (unsqueezed) Glauber coherent state describing the cavity photon wave, whose finite extent in both position and momentum quadratures is required by the Heisenberg uncertainty principle. If $Q_c < Q_a$, then the Lorentzian cavity transfer function acts a bandpass filter in which one can simultaneously look for signal power excess in Q_a/Q_c putative signal bins of width

$$\Delta f_s = \frac{m_a}{2\pi Q_a} \tag{12}$$

chosen to match the axion linewidth. This detection bandwidth is resolved by integrating for a time $\tau_a = 2\pi Q_a/m_a$ before performing the Fourier transform. The fluctuations in the thermal noise photon rate are then

$$\sigma_n = \sqrt{2}\frac{kT_{\text{sys}}}{\hbar m_a} \times \Delta f_s = \sqrt{2}\frac{kT_{\text{sys}}}{\hbar m_a}\frac{m_a}{2\pi Q_a} \tag{13}$$

where the system noise temperature $T_{\text{sys}} = \hbar m_a / k$ for a quantum-limited amplifier.

For a fixed total experimental duration t_{tot} of around 10^8 s to cover one octave in frequency, the total time that can be allocated for a particular cavity tuning is

$$t_{\text{tune}} = t_{\text{tot}}/Q_c \tag{14}$$

because Q_c steps of size $\Delta f_c = f_c/Q_c$ are required to cover an octave.

The measurements can be averaged over $N_{\text{avg}} = 2\Delta f_s t_{\text{tune}}$ measurements, each of duration $1/(2\Delta f_s)$ to obtain

$$\text{SNR}_{\text{SQL},Qc\leq Qa} \equiv \frac{R_s}{\sigma_n}\sqrt{2\Delta f_s t_{\text{tune}}} \tag{15}$$

$$= \frac{R_s}{\Delta f_s}\sqrt{\Delta f_s t_{\text{tune}}} \tag{16}$$

$$\propto \frac{Q_c}{1/Q_a} \sqrt{\frac{1}{Q_a} \frac{1}{Q_c}} \qquad (17)$$

$$\propto \sqrt{Q_a Q_c} \qquad (18)$$

2 Measurements in the Fock Basis Using Single Photon Resolving Detectors

Next, suppose that a single photon detector (SPD) is used to measure the cavity state in the Fock basis, with no sensitivity to the phase of the signal wave. This approach amounts to squeezing in the amplitude quadrature and can reduce the zero-point amplitude noise associated with SQL measurements in the Glauber coherent state basis utilized in the quantum-limited amplifiers considered above. This photon counter is read out once per cavity lifetime because, lacking the phase sensitivity, it cannot resolve frequency subcomponents within the cavity linewidth. For the tiny expected signal photon rates, the number of signal photons observed per cavity life-time will occupy a Poisson distribution with mean far less than 1 photon. So the photon counter will return a value of 0 or 1, and the distribution of the measurement values taken over many repeated measurements can be used to determine the Poisson mean of the signal. Suppose the dark count probability per readout is p_{err} and contributes photon counts due to spurious readout errors where the detector reports the presence of a cavity photon when no photon is actually present, or due to sensing the thermal photon population from an insufficiently cooled cavity. The rate of these background counts is

$$R_b = p_{err} \Delta f_c = p_{err} f / Q_c. \qquad (19)$$

(Note that $p_{err} = 1$ gives a value coinciding with the SQL noise fluctuations; the zero-point variance is equivalent to 1 background photon per resolved mode.) After an integration time t_{tune} at a fixed cavity tune, if the statistical uncertainties are dominated by the Poisson shot noise of background photons, the signal-to-noise ratio is

$$\text{SNR}_{\text{SPD},Q_c \leq Q_a} = \frac{R_s t_{tune}}{\sqrt{R_b t_{tune}}} \qquad (20)$$

$$= \frac{R_s t_{tune}}{\sqrt{p_{err} \Delta f_c t_{tune}}} \qquad (21)$$

$$= \frac{R_s}{\Delta f_s} \sqrt{\Delta f_s t_{tune}} \times \sqrt{\frac{\Delta f_s}{p_{err} \Delta f_c}} \qquad (22)$$

$$\propto \sqrt{Q_a Q_c} \times \sqrt{\frac{Q_c/Q_a}{p_{err}}} \qquad (23)$$

$$\propto \frac{Q_c}{\sqrt{p_{err}}} \tag{24}$$

An improvement over the sensitivity of the SQL amplifier Eq. 18 can, therefore, be achieved [3] if

$$p_{err} < Q_c/Q_a. \tag{25}$$

At $f = 10\,\mathrm{GHz}$, $Q_c \approx 10^4$ for copper cavities compatible with high magnetic fields, and so $p_{err} < 10^{-2}$ will enable an improved SNR, averaged over many measurements of duration equal to the cavity lifetime.

Quantum non-demolition (QND) detectors based on atoms or artificial atoms are a convenient approach to single microwave photon detection at microwave frequencies. Recall that the Lamb shift of a real atom is due to the motion of the electron relative to the nucleus as it is buffeted by the zero-point photon fluctuations. The inward and outward fluctuations of the electron orbit give asymmetric contributions to the average potential energy due to the nonlinearity of the Coulomb potential, and the average effect gives a net shift in the atomic energy levels. For sufficiently strong cavity enhancement, the electric field of single photons can also be measured via their additional contribution to the vacuum Lamb shift. The corresponding shift in the atomic transition frequencies—the AC Stark shift is quantized and proportional to the number of photons present in the cavity mode and so the atom can be used as a photon number resolving sensor. Just as with the Lamb shift, there is no net absorption of photons in this virtual process. So this nondestructive photon counting measurement can be repeated many times in order to achieve high measurement fidelity and low p_{err} [4].

The frequency shift of the atom (and the reciprocal frequency shift of the cavity) can be seen in the Jaynes–Cummings Hamiltonian

$$H_{JC} = \omega_c a^\dagger a + \omega_a \frac{\sigma_z}{2} + \frac{2g^2}{\omega_c - \omega_a} a^\dagger a \sigma_z + \cdots . \tag{26}$$

Here the nonlinearity of the atom allows it to be treated as a two-level system whose Fock representation can be modeled with spin-1/2 Pauli matrices. The third term is the interaction term which includes the dipole transition frequency g whose rate is filtered by the detuning of the cavity and atom frequencies. This large detuning makes the interaction virtual and suppresses the on shell absorption of photons by the atom. The resulting diagonalized interaction term is the electric polarizability of the atom which is bilinear in the number operators $a^\dagger a$ and σ_z. So a finite cavity photon occupation will shift ω_a by a quantized amount, and similarly, an atom excitation will shift ω_c. Since the interaction commutes with the bare Hamiltonian, the measurement can be QND. The required quantum back action is confined to the phase quadrature of the cavity photon mode. This phase noise arises because the spectroscopy necessarily involves the atom absorbing a probe photon at its shifted transition frequency, and this "spin flip" causes a reciprocal shift of the frequency of the coupled cavity mode.

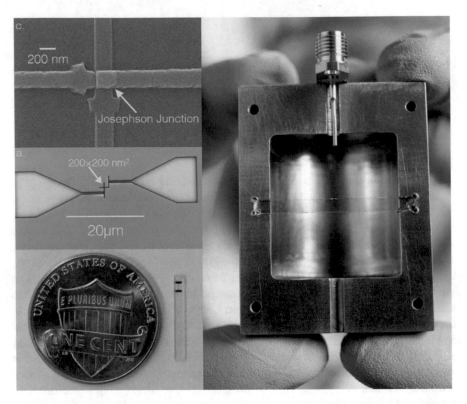

Fig. 1 Left: a qubit constructed by Akash Dixit (U.Chicago). The qubit is an anharmonic oscillator based on the nonlinear inductance of the Josephson junction. Its dipole coupling to the electric field of a cavity photon can be easily enhanced by attaching mm-scale antennae to opposite sides of the capacitive junction. Right: the electric field of even a single photon in a cavity can exercise the nonlinearity of the qubit oscillator (mounted inside the cavity with a glass slide) to produce a resolvable frequency shift. Spectroscopy of the qubit is performed with a simple antenna with feedthrough at the top of the cavity. Photo credit: Reidar Hahn (Fermilab)

Sufficiently low values of $p_{err} \sim 10^{-2}$ are routinely obtained in QND measurements utilizing superconducting qubits—anharmonic LC oscillators in which the cosine potential of a Josephson junction creates an inductance which is nonlinear in the Josephson tunneling current [5, 6]. Just as with atoms, the anharmonic qubit oscillator nondestructively encodes the electric field amplitude of the cavity photon as a shift of the qubit's own energy levels. Spectroscopy of the qubit's shifted transition frequency can then be used to determine the cavity mode's occupation number. These solid state devices are much easier to work with than real atoms and can be simply mounted in the cavity (Fig. 1). Spectroscopy is performed by scattering probe waves on the qubit, with the probe waves emitted and received by a simple antenna. R&D is underway at Fermilab (Aaron Chou/Daniel Bowring) and the University of Chicago (David Schuster) to further reduce the background rates in axion searches to levels well below the SQL.

3 Stimulated Emission of Axion Dark Matter into Photons

The axion wave and the cavity photon wave can be viewed as two coupled oscillators in which the transfer of energy is limited by the coherence times of the two oscillators. Classically, the power transfer can be enhanced if the cavity oscillator already has a nonzero amplitude: Power $= \overrightarrow{\text{force}} \cdot \overrightarrow{\text{velocity}}$ where the velocity of a harmonic oscillator is proportional to its amplitude. In quantum mechanics, the enhancement of transition rates due to finite amplitudes is encoded in the normalizations $a^\dagger|N\rangle = \sqrt{N+1}|N+1\rangle, a|N\rangle = \sqrt{N}|N-1\rangle$, and $a^\dagger|\alpha\rangle \approx a|\alpha\rangle = \alpha|\alpha\rangle$ of the Fock creation/annihilation operators acting on Fock states of amplitude $\sqrt{N}$ or Glauber states of amplitude α. The corresponding processes are called stimulated emission/absorption.

The primary obstacle to implementing a stimulated transition scheme to detect the axion dark matter is that the instantaneous phase of the classical axion wave is unknown and it changes by one radian every coherence time Q_a/m_a. So while one could populate the cavity state with a classical Glauber sine wave, one would not know which phase to use in order to stimulate the transfer of energy from the axion wave to the photon wave. If the wrong phase is chosen, then the stimulated absorption process dominates and energy is instead transferred from the photon wave to the axion wave. Moreover, even if stimulated emission could be achieved to enhance the signal rate, the Poisson statistics of the Glauber coherent state increases the Poisson noise in the cavity photon population in order to maintain a fixed signal-to-noise ratio equal to that of the SQL single-photon sensitivity. This noise enhancement is required by quantum mechanics in order to respect the SQL in any phase-sensitive measurement.

The solution is to utilize a phase independent initial cavity state—the Fock state $|N\rangle$—which is a state of definite photon number N and maximally indefinite phase. While the Fock state retains the $\sqrt{N+1}$ and $\sqrt{N}$ enhancements in stimulated transition amplitudes, it has no intrinsic Poisson noise and is instead a delta function in mode occupation number. It is also equally sensitive to any phase of the incoming axion wave. One can view the Fock state as a coherent superposition of Glauber states of all possible phases [7]. The subcomponents which are in phase with the axion wave will exhibit an enhanced transition $|N\rangle \rightarrow |N+1\rangle$ while the subcomponents which are out of phase with the axion wave will exhibit the enhanced transition $|N\rangle \rightarrow |N-1\rangle$.

The signature of the axion wave is a fast smearing of the initial cavity state, a delta function in the occupation number basis, into its neighboring bins of increased or decreased photon number. This transition can be again measured with photon number resolving detectors such as a qubit acting as a QND sensor to measure the occupation number distribution of the cavity as mapped onto a spectroscopic distribution of qubit resonances. In fact, qubits can also be used as single-photon buckets used to populate the cavity one quantum at a time in order to construct a cavity Fock state of definite photon number [8]. Research efforts are underway at JILA/Colorado (Konrad Lehnert) to implement this Fock-enhanced axion detection to vastly increase the axion signal rate while maintaining the low background rates of the qubit sensor.

Acknowledgements The experimental work outlined here is being performed in collaboration with Ankur Agrawal, Daniel Bowring, Akash Dixit, Konrad Lehnert, Reina Maruyama, and David Schuster. This work has been funded by the Heising-Simons Foundation. This document was prepared using the resources of the Fermi National Accelerator Laboratory (Fermilab), a U.S. Department of Energy, Office of Science, HEP User Facility. Fermilab is managed by Fermi Research Alliance, LLC (FRA), acting under Contract No. DE-AC02-07CH11359.

References

1. P. Sikivie, Phys. Rev. Lett. **51**, 1415 (1983). https://doi.org/10.1103/PhysRevLett.51.1415
2. C.M. Caves, Phys. Rev. D **26**, 1817 (1982). https://doi.org/10.1103/PhysRevD.26.1817
3. S.K. Lamoreaux, K.A. van Bibber, K.W. Lehnert, G. Carosi, Phys. Rev. D **88**, 035020 (2013). https://doi.org/10.1103/PhysRevD.88.035020
4. S. Gleyzes, S. Kuhr, C. Guerlin, J. Bernu, S. Deléglise, U. Busk Hoff, M. Brune, J.M. Raimond, S. Haroche, Nature **446**, 297 (2007). https://doi.org/10.1038/nature05589
5. J. Koch, T.M. Yu, J. Gambetta, A.A. Houck, D.I. Scuster, J. Majer, A. Blais, M.H. Devoret, S.M. Girvin, R.J. Schoelkopf, Phys. Rev. A - At. Mol. Opt. Phys. **76**(4), 1 (2007). https://doi.org/10.1103/PhysRevA.76.042319
6. B.R. Johnson, M.D. Reed, A.A. Houck, D.I. Schuster, L.S. Bishop, E. Ginossar, J.M. Gambetta, L. Dicarlo, L. Frunzio, S.M. Girvin, R.J. Schoelkopf, Nat. Phys. **6**(9), 663 (2010). https://doi.org/10.1038/nphys1710
7. B.C. Sanders, S.D. Bartlett, T. Rudolph, P.L. Knight, Phys. Rev. A **68**, 042329 (2003). https://doi.org/10.1103/PhysRevA.68.042329
8. M. Hofheinz, E.M. Weig, M. Ansmann, R.C. Bialczak, E. Lucero, M. Neeley, A.D. O'Connell, H. Wang, J.M. Martinis, A.N. Cleland, Nature **454**(7202), 310 (2008). https://doi.org/10.1038/nature07136

Light Dark Matter Searches at Accelerators and the LDMX Experiment

Bertrand Echenard

Abstract Accelerator-based experiments are playing an increasingly essential role in exploring the nature of dark matter. Several approaches have been proposed to search for light dark matter at collider and beam-dump experiments, providing unique sensitivity to several well-motivated scenarios. In this contribution, we review the current experimental situation and future efforts in that domain, emphasizing the advantages and challenges of each technique. A new proposal offering unprecedented sensitivity to directly annihilating thermal dark matter, the LDMX experiment, is also presented.

Keywords Light dark matter · Accelerators · Missing momentum · LDMX

1 Introduction

Collider and beam-dump experiments are increasingly recognized as indispensable tools in exploring dark matter (DM) in the vicinity of the known matter scales. Recent theoretical developments have motivated a large number of new ideas, a significant fraction of which could be explored in the near future. Among those models, thermal DM consisting of a relic whose density is set from nongravitational interactions with the standard model (SM) stands as particularly well-motivated. This scenario only requires that the DM-SM interaction rate exceeds the Hubble expansion in the early Universe for DM to thermalize, a rather generic condition. Cosmological constraints also restrict the mass of viable thermal DM to the keV–TeV range, a scale suggested by familiar matter.

Weakly interacting massive particles (WIMPs), an elegant realization of this paradigm, cover roughly the GeV–TeV range and have driven the experimental searches for the last decades. While WIMPs remain a well-motivated possibility, the simplest scenarios are becoming increasingly constrained. Less extensively studied, light DM spans the MeV–GeV region, and can be viewed as a paradigm where

B. Echenard (✉)
California Institute of Technology, Pasadena, CA 91125, USA
e-mail: echenard@caltech.edu

© Springer Nature Switzerland AG 2019
R. Essig et al. (eds.), *Illuminating Dark Matter*, Astrophysics and Space
Science Proceedings 56, https://doi.org/10.1007/978-3-030-31593-1_6

DM need not be tied strongly to Electroweak Symmetry breaking. Such a possibility arises naturally if the DM resides in a dark sector (DS) that interacts only feebly with the SM through a new set of interactions [1–3]. Such sectors are common in extensions of the SM, and a new force would extend the characteristics of thermal DM over the MeV–GeV range. Moreover, minimal DS models tend to exhibit a large degree of predictiveness, another attractive feature.

Dark matter annihilation leading to thermal equilibrium can only proceed through a few generic scenarios, depending on the DM and mediator (MED) mass. In the regime $m_{MED} < m_{DM}$, dark matter annihilates into DS particles, without any contact with the SM. The secluded annihilation rate, governed by the DM-mediator coupling in the DS, can be compatible with thermalization over a wide range of values [4, 5]. On the other hand, direct annihilation into SM particles occurs when $m_{MED} > m_{DM}$, and provides a clear, well-defined target. In that regime, the rate scales as

$$\langle \sigma v \rangle_{\text{direct}} \sim \frac{g_D^2 g_{SM}^2 m_{DM}^2}{m_{MED}^4}. \tag{1}$$

Since the dark sector coupling constant g_D (assuming perturbativity) and mass ratio m_{DM}/m_{MED} are at most $\mathcal{O}(1)$, the SM-mediator coupling g_{SM} must be above a certain threshold to be compatible with a thermal history. In other words, the dimensionless combination y must satisfy:

$$y \equiv \frac{g_D^2 g_{SM}^2}{16\pi^2} \left(\frac{m_{DM}}{m_{MED}} \right)^4 > \langle \sigma v \rangle_{\text{relic}} \; m_{DM}^2 \tag{2}$$

which is qualitatively valid regardless of the DM nature. The lower bound defines a predictive target

$$y_{\text{target}} \equiv g_D^2 \, g_{SM}^2 \left(\frac{m_{DM}}{m_{MED}} \right)^4 \tag{3}$$

as a function of the DM mass to achieve thermalization with the SM. Larger values correspond to models where direct annihilation is only a subdominant process determining the DM abundance.

While the argument, so far, is applicable to any type of interactions between the SM and the DS, the vector/kinetic mixing portal is by far the most viable among renormalizable operators [3, 6]. In the most popular scenario, the interaction between the DS and the SM is mediated by a dark photon (A') with a dark photon–photon mixing strength ϵ. Variations on this theme include models in which the mediator couples preferentially to baryonic (leptophobic DM), leptonic (leptophilic DM), or $(B - L)$ currents. Dark matter annihilation on the CMB power spectrum provides important constraints on the vector portal (see for example Ref. [7]), ruling out direct annihilation of Dirac fermions. The remaining possibilities experience reduced annihilation due to velocity suppression (scalar and Majorana DM) or population suppression if the leading annihilation involves an excited state (pseudo-Dirac DM).

Besides directly annihilating thermal DM, the experimental approaches discussed below are also sensitive to a broad array of models. Those include secluded thermal DM, asymmetric DM in which the DM abundance arises from a primordial asymmetry [8]; SIMP DM containing new resonances in the DS [9]; models with different cosmological histories, such as ELDER DM [10]; freeze-in models with heavy mediators [11, 12]; new force carriers decaying to SM particles [2] or searches for millicharged particles [13, 14].

In the following, we'll briefly review the different techniques to search for DM at accelerators, with a focus on directly annihilating thermal DM. Colliders and fixed-target experiments have already explored a large portion of the parameter space, and they are poised to make significant progress in the coming decade. The description of a new proposal to search for DM, the LDMX experiment, will close the discussion.

2 Dark Matter Searches at Accelerators

Compared to other approaches, fixed target and collider experiments offer several key advantages. Relativistic DM production is largely independent of the details of the DS, as illustrated in Fig. 1. In some models, e.g., Majorana fermion DM interacting through a vector, the direct detection cross section σ_{DD} is drastically reduced through its dependence on the DM velocity. On the other hand, DM particles are produced relativistically at accelerators, and the scattering cross section is only

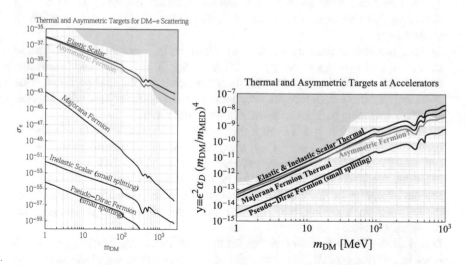

Fig. 1 Targets for directly annihilating thermal DM and asymmetric DM for (left) nonrelativistic electron-DM scattering probed by direct detection experiments and (right) relativistic accelerator-based experiments. The various lines depict the scalar, Majorana, inelastic, and pseudo-Dirac DM annihilating through the vector portal. Current constraints are shown in gray-shaded areas. Figures taken from Ref. [3]

weakly dependent on the velocity. In the pseudo-Dirac case, DM (now labeled χ_1) is accompanied by a heavier state (χ_2), and annihilation or scattering can proceed dominantly via off-diagonal couplings between the light mediator and the χ_1 and χ_2 particles. Direct detection scattering can be heavily suppressed when the DM kinetic energy is insufficient to produce the heavier state [15]. In contrast, the lighter state can readily up-scatter to the excited state when scattering off a nucleus.

Besides detection capabilities, accelerator-based experiments also offer a way to study the DS structure and determine the parameters of a Lagrangian. The mass of the mediator could be measured with visible SM decays, as well as specific type of reactions for invisible decays. The nature of the mediator-SM coupling, another fundamental property, could be investigated using proton (quark coupling) or electron (leptonic coupling) beams. Experiments detecting DM by its scattering in a target would also provide insights about the DS coupling constant.

While accelerator-based approaches have many advantages, some possibilities remain only accessible to direct detection experiments, such as freeze-in models with an ultralight mediator or ultralight bosonic DM. Direct detection would also be desirable to establish the cosmological nature of any observation. A multipronged approach would therefore be advocated to explore as much parameter space as possible and untangle the physics of the dark sector.

On the experimental side, several techniques have been proposed to search for DM signatures. As a broad organizing principle, they can be classified as follows:

- **Missing mass**: the DM signature is identified as a resonance in the recoil mass distribution against a fully reconstructed final state, for example, $e^+e^- \rightarrow \gamma(A' \rightarrow \chi\bar{\chi})$ annihilations. As all particles (including initial ones) but the DM must be well measured, this type of search is usually performed at e^+e^- colliders or positron beam dumps.
- **Missing momentum/energy**: the DM is radiated off the incoming electron/proton in $eZ \rightarrow eZ(A' \rightarrow \chi\bar{\chi})$ or $pp \rightarrow X(A' \rightarrow \chi\bar{\chi}\ X = \gamma, \text{jet})$ and identified through the missing energy/momentum carried away by the DM particles. This approach requires a detector with excellent hermicity, and the possibility to measure each incoming particle separately in some instances.
- **Electron and proton beam dump**: the production mechanism relies on meson decays, such as $\pi^0/\eta^{(')} \rightarrow \gamma(A' \rightarrow \chi\bar{\chi})$, or radiation off electrons ($eZ \rightarrow eZ(A' \rightarrow \chi\bar{\chi})$) or protons ($pe(p) \rightarrow pe(p)A', A' \rightarrow \chi\bar{\chi}$). The DM is usually detected in a downstream detector via $e\chi \rightarrow e\chi$ or $N\chi \rightarrow N\chi$ scattering. This technique has the advantage or probing the DM interaction twice, providing sensitivity to the DS-mediator coupling, but requires a large incoming flux to compensate for the reduced yields.
- **Direct dark photon searches**: search for the mediator through its decays into SM particles. This approach is essential for $m_{A'} < 2m_\chi$, when the mediator decays visibly. Many production mechanisms are possible, e.g., $e^+e^- \rightarrow \gamma A'$, $eZ \rightarrow eZA'$ or neutral meson decays. The mediator is usually reconstructed though its leptonic decays.

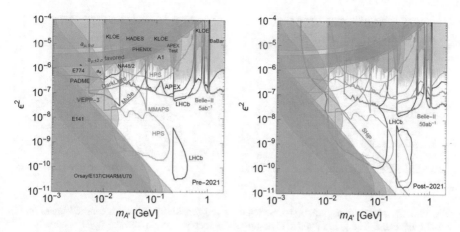

Fig. 2 Existing constraints on visibly decaying dark photons (shaded regions) and projected sensitivities of future and proposed experiments (solid lines). Visible decays of the mediator dominate in the secluded annihilation regime. Courtesy R. Essig

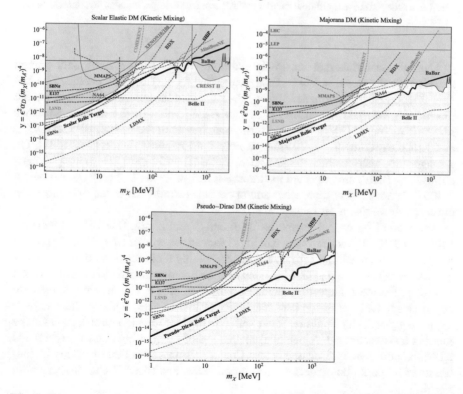

Fig. 3 Current constraints (shaded regions) and sensitivity estimates (dashed/solid lines) on the parameter y for (top left) elastic DM, (top right) Majorana DM, and (bottom) pseudo-Dirac DM. The calculations are performed using $m_{A'} = 3m_\chi$ and $\alpha_D = 0.5$, *conservative values* of the parameters. For larger ratios or smaller values of α_D, the accelerator-based experimental curves shift downward, but the thermal relic target remains invariant. Courtesy G. Krnjaic

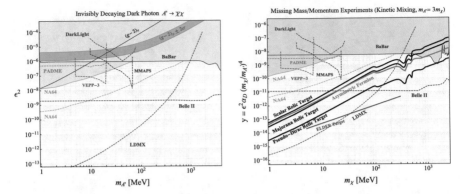

Fig. 4 Left: current constraints (shaded regions) and sensitivity estimates (dashed lines) on the kinetic mixing ϵ. The green band shows the values required to explain the (g-2)$_\mu$ anomaly [19]. Right: corresponding curves on the parameter y, together with the asymmetric DM and ELDER targets (orange and magenta lines). The calculations are performed sing $m_{A'} = 3m_\chi$ and $\alpha_D = 0.5$, *conservative values* of the parameters. For larger ratios or smaller values of α_D, the accelerator-based experimental curves shift downward, but the thermal relic target remains invariant. Courtesy G. Krnjaic

Current constraints and sensitivity estimates for visibly decaying dark photon searches are displayed in Fig. 2. Past measurements have already excluded a sizeable fraction of the parameter space [16–18], including values suggested by the discrepancy between the measured and predicted value of the muon anomalous magnetic moment [19]. In the short term, searches from the APEX, HPS, PADME, and LHCb experiments will further explore the low-mass region [2, 20, 21]. On a longer timescale, collider (e.g., LHCb, Belle-II) and future beam-dump experiments (e.g., SHiP) are projected to almost entirely probe dark photon masses below ∼400–500 MeV. New approaches and/or facilities would be needed to improve the coverage above that mass range.

The present status and prospects for directly annihilating DM with a kinematically mixed dark photon are shown in Fig. 3 for various type of DM. While important progress has been achieved from searches at existing facilities or reinterpretation of previous results (see, e.g., [22, 23]), a next generation of experiment is clearly needed to explore the most interesting region of parameter space. The missing momentum approach seems to offer the best sensitivity at low masses, while collider experiments (e.g., Belle-II) would be better suited to explore the high mass region via the missing mass technique. A potential realization of the missing mass approach, the LDMX experiment, is discussed below. Constraints on a few other scenarios, including invisible dark photon decays, asymmetric DM, and ELDER DM, are shown in Fig. 4.

3 The LDMX Experiment

The "Light DarkMatter eXperiment" (LDMX) [24] aims to precisely measure missing momentum and energy in electro-nuclear collisions in a thin target with unprecedented sensitivity. To achieve high statistics, LDMX plans to use a low current, high-repetition electron beam with a 4–10 GeV energy. In the first phase, LDMX would collect a sample of 4×10^{14} electrons on target (EOT) at a rate of 10^8 electrons per second ($\sim 1\,e^-$ per bunch), before increasing the sample size by two orders of magnitude in Phase-II. The proposed DASEL beam-line at SLAC [25], CEBAF at Jefferson Lab, or a new beam-line at CERN [26] are potential candidate to host this experiment. Beside dark matter, electro-nuclear and photo-nuclear reactions of broader interest to the neutrino community could be also studied with LDMX.

The kinematics of every incident electron is reconstructed both up- and downstream of the target by a tracking system placed into a weak magnetic field, while additional neutral activity is detected by electromagnetic (ECAL) and hadronic (HCAL) calorimeters downstream of the target, with a sensitive area extending onto the beam axis itself. The upstream tracker will reject with very high efficiency stray low-energy particles from the beam halo that could mimic the DM signal. These four detector systems: the upstream tagging tracker, the downstream recoil tracker, the forward ECAL, and HCAL hadronic calorimeter form the majority of the LDMX experimental concept. To keep the detector compact and the field in the ECAL minimal, the tagging tracker is placed inside the bore of a dipole magnet and the recoil tracker in its fringe field. This layout is illustrated in Fig. 5.

The tracker and calorimeters must be able to contend with a high rate of events producing one of the several dominant topologies. Electrons might not interact significantly in the target, resulting in a hard track through both trackers and an energetic shower in the ECAL. Electrons could also emit an energetic photon while interacting in the target. These "hard bremsstrahlung" topologies feature a low-energy recoil electron similar to signal electrons and two showers in the ECAL, with large combined shower energy, separated by a few cm. Finally, trident events contain two or three tracks reconstructed by the tracker (depending on kinematics) and several ECAL showers. In addition, the calorimeters must veto with extreme efficiency a wide variety of sub-dominant backgrounds, such as a hard bremsstrahlung photon undergoing a photo-nuclear reaction producing only a few energetic ($\mathcal{O}(1\,\text{GeV})$) neutrons escaping from the nucleus.

These considerations call for a fast, high-precision tracking system and a high-speed, high-granularity Silicon calorimeter, used in conjunction with a hadron calorimeter to achieve the desired level of rejection. The LDMX concept plans to meet these challenges by leveraging technology under development for the silicon sampling calorimeter for the CMS high luminosity forward calorimeter upgrade [27], and the tracking technology developed for the HPS experiment [28].

The sensitivity of the LDMX experiment is shown in Fig. 3 for the thermal relic DM scenarios described previously. LDMX will have unprecedented sensitivity surpassing all existing and projected constraints by orders of magnitude for DM masses

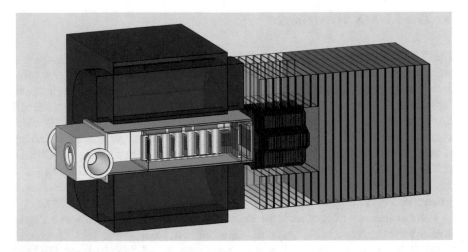

Fig. 5 A cutaway overview of a potential LDMX detector design showing, from left to right, the trackers and target in the spectrometer dipole, the forward electromagnetic, and hadronic calorimeters. The final design is still under study. Courtesy T. Nelson

below a few hundred MeV. LDMX aims in its first phase to fully explore the scalar and Majorana fermion thermal DM parameter space in that mass range, and the remaining possibilities in its second phase. The experiment will also greatly improve the sensitivity to invisible dark photon decays, asymmetric DM, ELDER/SIMP scenarios and light long-lived neutral particles.

Acknowledgements We would like to thank the Simons foundation and the organizers of the symposium. The author is indebted to N. Toro, R. Essig, and G. Krnjaic for their useful discussion and providing several figures. This work has also benefited from the many contributions of the members of the LDMX Collaboration. The author is supported by the US Department of Energy under grant DE-SC0011925.

References

1. C. Boehm, P. Fayet, Nucl. Phys. B **683**, 219 (2004). https://doi.org/10.1016/j.nuclphysb.2004.01.015
2. J. Alexander et al. (2016), http://inspirehep.net/record/1484628/files/arXiv:1608.08632.pdf
3. M. Battaglieri et al. (2017)
4. D.P. Finkbeiner, N. Weiner, Phys. Rev. D **76**, 083519 (2007). https://doi.org/10.1103/PhysRevD.76.083519
5. M. Pospelov, A. Ritz, M.B. Voloshin, Phys. Lett. B **662**, 53 (2008). https://doi.org/10.1016/j.physletb.2008.02.052
6. G. Krnjaic (2015)
7. D.P. Finkbeiner, S. Galli, T. Lin, T.R. Slatyer, Phys. Rev. D **85**, 043522 (2012). https://doi.org/10.1103/PhysRevD.85.043522
8. K.M. Zurek, Phys. Rep. **537**, 91 (2014). https://doi.org/10.1016/j.physrep.2013.12.001

9. Y. Hochberg, E. Kuflik, H. Murayama, JHEP **05**, 090 (2016). https://doi.org/10.1007/ JHEP05(2016)090
10. E. Kuflik, M. Perelstein, N.R.L. Lorier, Y.D. Tsai, Phys. Rev. Lett. **116**(22), 221302 (2016). https://doi.org/10.1103/PhysRevLett.116.221302
11. L.J. Hall, K. Jedamzik, J. March-Russell, S.M. West, JHEP **03**, 080 (2010). https://doi.org/10. 1007/JHEP03(2010)080
12. A. Berlin, N. Blinov, G. Krnjaic, P. Schuster, N. Toro (2018)
13. A.A. Prinz et al., Phys. Rev. Lett. **81**, 1175 (1998). https://doi.org/10.1103/PhysRevLett.81. 1175
14. A. Haas, C.S. Hill, E. Izaguirre, I. Yavin, Phys. Lett. B **746**, 117 (2015). https://doi.org/10. 1016/j.physletb.2015.04.062
15. D. Tucker-Smith, N. Weiner, Phys. Rev. D **64**, 043502 (2001). https://doi.org/10.1103/ PhysRevD.64.043502
16. J.P. Lees et al., Phys. Rev. Lett. **113**(20), 201801 (2014). https://doi.org/10.1103/PhysRevLett. 113.201801
17. R. Aaij et al., Phys. Rev. Lett. **120**(6), 061801 (2018). https://doi.org/10.1103/PhysRevLett. 120.061801
18. J.R. Batley et al., Phys. Lett. B **746**, 178 (2015). https://doi.org/10.1016/j.physletb.2015.04. 068
19. M. Pospelov, Phys. Rev. D **80**, 095002 (2009). https://doi.org/10.1103/PhysRevD.80.095002
20. P. Ilten, J. Thaler, M. Williams, W. Xue, Phys. Rev. D **92**(11), 115017 (2015). https://doi.org/ 10.1103/PhysRevD.92.115017
21. P. Ilten, Y. Soreq, J. Thaler, M. Williams, W. Xue, Phys. Rev. Lett. **116**(25), 251803 (2016). https://doi.org/10.1103/PhysRevLett.116.251803
22. J.P. Lees et al., Phys. Rev. Lett. **119**(13), 131804 (2017). https://doi.org/10.1103/PhysRevLett. 119.131804
23. D. Banerjee et al., Phys. Rev. D **97**(7), 072002 (2018). https://doi.org/10.1103/PhysRevD.97. 072002
24. https://confluence.slac.stanford.edu/display/MME
25. T. Raubenheimer, A. Beukers, A. Fry, C. Hast, T. Markiewicz, Y. Nosochkov, N. Phinney, P. Schuster, N. Toro, *DASEL: Dark Sector Experiments at LCLS-II* (2018). arxiv: 1801.07867
26. T. Akesson, Y. Dutheil, L. Evans, A. Grudiev, Y. Papaphilippou, S. Stapnes, *A Primary Electron Beam Facility at CERN* (2018). arxiv: 1805.12379
27. D. Contardo, M. Klute, J. Mans, L. Silvestris, J. Butler, *Technical Proposal for the Phase-II Upgrade of the CMS Detector* (2015)
28. M. Battaglieri et al., Nucl. Instrum. Methods A **777**, 91 (2015). https://doi.org/10.1016/j.nima. 2014.12.017

Direct Detection of Sub-GeV Dark Matter: Models and Constraints

Rouven Essig

Abstract I will make some general comments about the search for dark matter and other new particles, contrasting current research trends with those 10 years ago. I will then focus on recent ideas for direct detection experiments to search for dark matter with masses in the MeV-to-GeV range. I will then discuss briefly three topics: (i) the solar neutrino background (or "how low in cross section (interaction strength) can future direct-detection experiments probe before solar neutrinos become an irreducible background"), (ii) novel constraints on low-mass dark matter from Supernova 1987A, and (iii) strongly interacting dark matter (or "how large in cross section can direct-detection experiments probe before terrestrial effects stop sub-GeV dark matter from reaching the detector").

1 A Personal Perspective

Ten years ago, I was transitioning from completing my Ph.D. degree to starting a postdoctoral research position. It was an exciting time to be a particle physicist in the pursuit of physics beyond the Standard Model! The LHC was about to turn on in September 10, 2008, with the potential of discovering new particles at the Weak scale that would solve the Higgs hierarchy problem and with the potential of discovering dark matter. There seemed to be anomalies related to dark matter everywhere. The satellite experiments PAMELA and ATIC saw an exciting excess of cosmic-ray positrons, well above the expected signal from astrophysical processes, hinting at the existence of dark matter annihilating into leptons. WMAP, EGRET, and INTEGRAL data hinted at the existence of dark matter. Data from dark matter direct detection experiments, especially DAMA/NaI and DAMA/LIBRA, hinted at a dark matter signal, and several other direct detection experiments were taking data, or about to take data. The Fermi Gamma-ray Space Telescope was just launched, in

R. Essig (✉)
C.N. Yang Institute for Theoretical Physics, Stony Brook University,
Stony Brook, NY 11794, USA
e-mail: rouven.essig@stonybrook.edu
URL: http://insti.physics.sunysb.edu/~rouven

© Springer Nature Switzerland AG 2019
R. Essig et al. (eds.), *Illuminating Dark Matter*, Astrophysics and Space
Science Proceedings 56, https://doi.org/10.1007/978-3-030-31593-1_7

June 2008, with the potential of discovering dark matter annihilating or decaying to Standard Model particles. Many collider, terrestrial, and satellite experiments were poised to take new data. It was exciting both because of anomalies in existing data and because of the upcoming deluge of data. The expectation that we would discover several new fundamental building blocks of Nature and one or more mediators of new interactions was high—and entirely reasonable.

Ten years later, our expectations of discovering dark matter and other new particles beyond the Standard Model have not yet been met.[1] The anomalies related to dark matter from a decade ago have either disappeared or currently have a low probability of having a dark matter origin. The deluge of data has come without many hints for new physics. While a few interesting anomalies remain and may prove to be new physics, I think that in the community as a whole these currently do not rise to the same level of excitement as the anomalies 10 years ago.

Despite not having found new physics, this past decade has been a fantastic success in every other way. We have learned much about particle physics, the Standard Model, cosmology, and astrophysics, and despite not having met our high expectations from 10 years ago, the field of particle physics, and the search for dark matter and other physics beyond the Standard Model, is healthy, vibrant, and exciting. Moreover, the coming decade will still be amazingly data rich. "Big" experiments like the LHC will keep probing for dark matter and other new particles. Other collider, terrestrial, and satellite experiments are, or will soon, take data. While our hope of finding new physics beyond the Standard Model in the coming decade may be slightly lower overall today than it was ten years ago, the expectation that we will discover dark matter or some other new physics in the next few years is still entirely reasonable.

A major reason for my excitement, and one piece of evidence for a healthy field, is that particle physicists have come up with many ideas for new *small-scale* experiments that will allow us yet again to probe orders of magnitude of uncharted dark matter (and other) parameter space over the next decade (see, for example, [1–4]). This effort spans across several, traditionally disparate, disciplines. Indeed, particle theorists have been working together with particle experimentalists, instrumentalists, condensed matter theorists, AMO theorists, AMO experimentalists, and others to develop new ideas, new techniques, and new detectors for a new generation of experiments. These experiments can probe an array of "non-traditional" (non-WIMP) dark matter candidates over a mass range that spans perhaps 30 orders of magnitude below the Weak scale. While upcoming WIMP searches are very important and remain well-motivated, a much broader approach to finding dark matter is necessary. Several experimental ideas to search for new force mediators or for dark matter consisting of axion-like particles or dark photons, or for dark matter with MeV-to-GeV masses have been realized, or will be realized soon, but more funding is needed to support this still-developing field. I am optimistic that this funding is forthcoming,

---

[1] The discovery at the LHC of the Higgs boson is a notable success, and while the current lack of evidence for other particles at the Weak scale has sharpened the hierarchy problem, the Higgs boson currently conforms to the Standard Model expectations.

since the science case for these new searches is strong. Moreover, even a small, inexpensive experiment can probe vast regions of motivated and unexplored parameter space by exploiting novel detector technologies.

2   Direct Detection of Sub-GeV Dark Matter

In my talk at the Simons Symposium, I focused on the direct detection of MeV-to-GeV mass dark matter. There are many viable candidates for dark matter in this mass range, which can obtain the observed relic abundance from various possible mechanisms (thermal, asymmetric, ELDER, SIMP), see [4] for a summary. In many cases, the simplest known possibilities have not yet been experimentally fully probed. Traditional WIMP detection relies on searching for nuclear recoils from WIMPs scattering off nuclei in the detector. However, the nuclear recoil energies from low-mass dark matter is typically below the detector thresholds (the lowest mass currently probed is $\sim$120 MeV [5]), so other techniques are needed. One promising avenue is to search for electron recoils from dark matter scattering off electrons [6]. This idea was proposed several years ago, but only recently are experiments reaching the required sensitivity while being able to control backgrounds.

The first limits on dark matter masses down to $\sim$4 MeV based on this idea were derived in [7], using data from XENON10 [8]. However, while XENON10, and later also XENON100 and DarkSide-50, have all demonstrated sensitivity to low-mass dark matter scattering off electrons [8–10], large detector-specific backgrounds are severely limiting the sensitivity and discovery potential of these experiments with noble liquid targets. Efforts are underway by the LBECA Collaboration [4] to mitigate these backgrounds and build a 10-kg xenon detector focused on electron recoils [11].

Other target materials and detector setups may be more successful. Indeed, two major recent technological successes stand out in the search for electron recoils from low-mass dark matter, one by the SENSEI Collaboration and one by the SuperCDMS Collaboration. (i) SENSEI—the Sub-Electron Noise Skipper-CCD Experimental Instrument—uses thick fully depleted silicon CCD ("Skipper CCDs") that have ultralow readout noise ($\sim$0.05 rms/pix). Individual electrons can be counted [12], and the detector does not suffer from similar backgrounds that currently plague the noble liquid detectors. The first dark matter search results, using a $\sim$0.1 g prototype Skipper CCD, were presented in [13]. The goal is to build an experiment using 100-grams of Skipper CCDs. I refer the reader to the contribution in these proceedings of my collaborator, Javier Tiffenberg, for more details. (ii) SuperCDMS, on the other hand, drifts electrons across a large bias voltage to amplify the small charge signal into a large phonon signal through the Neganov–Luke effect; the resulting phonons are measured with transition edge sensor technology. Individual electron can again be counted [14], and the first dark matter search results using a prototype $\sim$1 g detector were presented in [15].

With this background, I will now briefly summarize the results discussed in my talk, which focuses on three topics: (i) the solar neutrino background (or "how low in cross section (interaction strength) can future direct-detection experiments probe before solar neutrinos become an irreducible background"), (ii) novel constraints on low-mass dark matter from Supernova 1987A, and (iii) strongly interacting dark matter (or "how large in cross section can direct-detection experiments probe before terrestrial effects stop sub-GeV dark matter from reaching the detector").

3   Results

3.1   Solar Neutrino Background

The results in this section are based on [16].

Direct detection experiments searching for electron recoils will also be sensitive to solar neutrinos via coherent neutrino-nucleus scattering (CNS) [6], since the recoiling nucleus can produce a small ionization signal. (Note that solar neutrinos can also scatter directly off electrons, but the resulting electron recoils are typically at much higher energies than the electron recoil energies of interest from dark matter.) Even if one overcomes the challenges of controlling both radioactive and detector-specific backgrounds, solar neutrinos will eventually present a background to a dark matter search that cannot be controlled or reduced by improved shielding or material purification and handling. It is thus worth understanding how solar neutrinos would eventually limit the sensitivity of direct detection experiments sensitive only to electron recoils. This is a topic that has been discussed extensively for nuclear recoils, see, e.g., [18] and references therein.

We find that solar neutrinos begin limiting the sensitivity to dark matter electron scattering in silicon targets for exposures larger than about a few kg-years. In xenon targets, the CNS rate is higher, and it turns out that solar neutrinos already start limiting the sensitivity for exposures as little as a few hundred gram years. Said in another way, xenon detectors could be sensitive to solar neutrinos (especially, the ^{8}B component) even for relatively small exposures compared to silicon detectors. Figure 1 shows the results in more detail.

3.2   Constraints from SN1987A on Sub-GeV Dark Matter

The results in this section are based on [19].

A supernova—SN1987A—observed in 1987 in the Large Magellanic Cloud provides important constraints on the existence of sub-GeV dark matter and other low-mass dark sectors, since these particles could provide novel channels to "cool" the proto-neutron star and change the observed neutrino emission from SN1987A. Constraints are derived by requiring that the luminosity carried by the new particles from the interior of the proto-neutron star environment to the outside of the neutrinosphere is smaller than the luminosity carried by neutrinos [20].

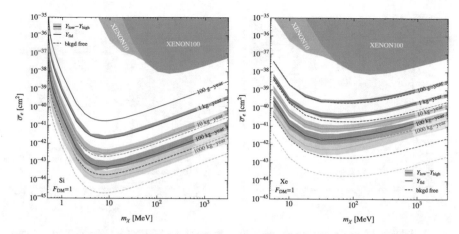

Fig. 1 Discovery limits for dark matter electron scattering in silicon (left) and xenon (right) assuming a the scattering is mediated by a "heavy" particle, leading to a momentum-independent dark matter form factor ($F_{DM} = 1$). The figures are taken from [16]. Exposures of 0.1, 1, 10, 100, and 1000 kg-years are shown in various colors. Since a significant uncertainty in estimating the solar neutrino background is how much ionization is generated by low-energy nuclear recoils, we calculate the discovery limits under different assumptions for the low-energy ionization efficiency. The solid line shows the results assuming a "fiducial" ionization efficiency (defined in [17]), while the shaded bands denote the range between a high and low ionization efficiencies. The dashed lines show the background-free 90% C.L. sensitivities. The gray shaded region shows the current direct detection limits on DM-electron scattering from [17]

In [19], we derive constraints from SN1987A on several possible low-mass particles: various dark sectors consisting of dark matter and dark photons (including millicharged particles), the QCD axion, and axion-like particles with Yukawa couplings. In my presentation, I focused on dark sectors consisting of dark matter and dark photons, both for "heavy" dark photons as well as for an ultralight dark photon (in which case bounds derived for millicharged dark matter particles are applicable). These are popular benchmark models when presenting prospects for new direct detection and accelerator-based searches for MeV-to-GeV mass dark matter [4]. The constraints on these dark sectors are shown in Fig. 2.

We see that our bounds are applicable for small couplings and masses $\lesssim 100$ MeV, and do not decouple for low fermion masses. They exclude parameter space that is otherwise unconstrained by existing accelerator-based and direct detection searches. Importantly, our bounds are complementary to proposed laboratory searches for sub-GeV dark matter, and do not constrain several benchmark model targets in parameter space for which the dark matter obtains the correct relic abundance from interactions with the Standard Model.

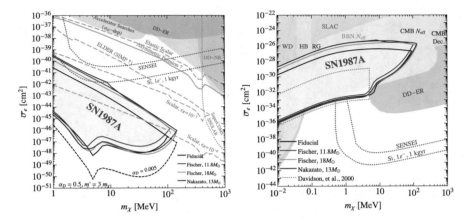

Fig. 2 Thick solid black, red, green, and red lines show the SN1987A constraints for various temperature and density profiles on "light" dark matter coupled to a dark photon, assuming the specific dark matter–dark photon mass relation $m' = 3m_\chi$, and $\alpha_D = 0.5$. Dashed black line shows the constraint for $\alpha_D = 0.005$ using the fiducial profile. Thick orange lines show several benchmark model "targets", along (or above) which the DM can obtain the correct relic abundance in various scenarios. Existing laboratory-based searches are shown in gray, including colliders, beam-dump, and fixed-target experiments that search for dark photons decaying to dark matter ($A' \rightarrow \chi\bar{\chi}$). **Left**: Under the assumption that the χ is all of the dark matter, we show constraints from dark matter electron scattering from XENON10, XENON100, and DarkSide-50, and constraints on dark matter nucleus scattering from the CRESST, SuperCDMS, and LUX collaborations. Dotted lines show projections SuperCDMS SNOLAB (green), as well as SENSEI and a hypothetical experiment using a silicon target sensitive to single electrons with a 1 kg-year exposure (both blue) [12, 21]. **Right**: Thick solid black, red, green, and red lines show the SN1987A constraints for various temperature and density profiles on dark matter coupled to an ultralight dark photon mediator. Other relevant bounds, derived specifically for millicharged particles but also relevant for an ultralight dark photon mediator, are taken from [22–24]. Our SN1987A bound updates a previous bound presented in [25] (dotted line). Plots are taken from [19], which contains additional references and details

3.3 Strongly Interacting Dark Matter

The results in this section are based on work in progress with Timon Emken, Christopher Kouvaris, and Mukul Sholapurkar [26].

Dark matter that interacts strongly with the visible sector cannot penetrate the Earth, so that detectors placed deep underground have no sensitivity. Detectors operating at shallow sites, on the surface, or even on balloons or satellites can probe parameter space that is unconstrained by underground detectors, despite large cosmic-ray backgrounds.

The terrestrial effects on MeV-to-GeV dark matter scattering off nuclei or electrons are model-dependent and have so far only been explored partially in the literature [27–29]. Let us consider a dark photon mediator that allows dark matter to scatter off nuclei and electrons in the atmosphere or Earth (we include elastic scatters only, ignoring inelastic scatters off electrons). The dark matter then reaches the detector, where it scatters off an *electron* to leave an observable signal.

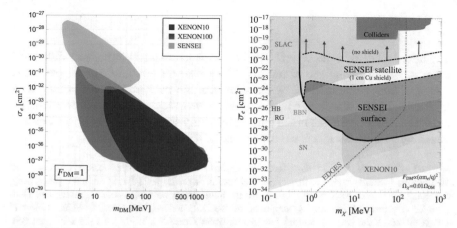

Fig. 3 **Left**: The red-, blue-, and green-shaded regions are the current constraints from XENON10 [7], XENON100 [17], and SENSEI [13], in which we include terrestrial effects for which dark matter interacting with a dark photon can be stopped in the Earth's crust or atmosphere for large cross sections. Both XENON10 and XENON100 are deep underground, beneath the Gran Sasso mountain, while the shown results from SENSEI is from a surface run. We assume the dark photon mediator is "heavy", so that the dark matter form factor is unity. Note that this plot is preliminary, and several other experiments are missing, notably DarkSide-50 [10] and SuperCDMS-HV [15]. **Right**: The constraints from XENON10 (light blue) and a SENSEI surface run (red) on dark matter interacting with an ultralight mediator. Other constraints include SN1987A (yellow region labeled "SN") [19], red-giant (red, "RG") and horizontal branch (brown, "HB") stars [22], and Big Bang Nucleosynthesis (green, "BBN") [22], and the SLAC millicharge experiment (gray, "SLAC") [23]. A Skipper-CCD on a satellite with 1 cm of copper shielding (black dot-dashed line, "SENSEI satellite") would extend the SENSEI surface run to even larger cross sections, and less shielding would of course probe even larger cross sections (arrows)

In this context, we can map out the parameter space constrained by several direct detection experiments sensitive to electron recoils. In Fig. 3, we consider both a heavy dark photon (left plot) and an ultralight dark photon (right plot). We see that the direct detection parameter space that is constrained by deep underground detectors (XENON10/100) is largely complementary to that constrained by a detector placed on the surface (like the SENSEI surface run).

We also explored the possibility of having an ultralow threshold detector placed on a satellite in Fig. 3 (right). This would allow us to probe even larger cross sections that the SENSEI surface run from [13]. This could be very interesting in the context of dark matter models (such as millicharged dark matter with a density of $\sim 1\%$) that could explain the EDGES measurement of the 21-cm spectrum at $z \simeq 17$, which revealed an anomalously large absorption signal [30, 31] (the green dashed line shows a line from [32] in which the EDGES anomaly could be explained). However, the viability of the open parameter space, and whether a SENSEI Skipper CCD on a satellite could probe it, is currently under investigation.

4 Outlook

The next decade will see several small-scale experiments being operated that will cover orders of magnitude of unexplored dark sector and dark matter parameter space, without being affected by the solar neutrino background or constrained by SN1987A. Moreover, placing a detector sensitive to low-energy electron recoils on a satellite may be essential in order to explore fully strongly interacting dark matter, including dark matter that could explain the EDGES signal.

Acknowledgements I would like to thank the Simons Foundation for their generous support of this symposium. I would also like to thank my collaborators on the projects discussed in this proceeding, Jae Hyeok Chang, Timon Emken, Chris Kouvaris, Sam McDermott, Mukul Sholapurkar, and Tien-Tien Yu. My research is currently supported by the DoE under Grant Nos. DE-SC0017938 and DE-SC0018952, the Heising-Simons Foundation under Grant No. 79921, and the US-Israel BSF under Grant No. 2016153.

References

1. *Fundamental Physics at the Intensity Frontier.* https://doi.org/10.2172/1042577, http://inspirehep.net/record/1114323/files/arXiv:1205.2671.pdf
2. R. Essig, et al., in *Proceedings, 2013 Community Summer Study on the Future of U.S. Particle Physics: Snowmass on the Mississippi (CSS2013): Minneapolis, MN, USA, July 29-August 6, 2013* (2013). http://inspirehep.net/record/1263039/files/arXiv:1311.0029.pdf
3. J. Alexander, et al., (2016). http://inspirehep.net/record/1484628/files/arXiv:1608.08632.pdf
4. M. Battaglieri, et al. (2017)
5. G. Angloher et al., Eur. Phys. J. C **77**(9), 637 (2017). https://doi.org/10.1140/epjc/s10052-017-5223-9
6. R. Essig, J. Mardon, T. Volansky, Phys. Rev. D **85**, 076007 (2012). https://doi.org/10.1103/PhysRevD.85.076007
7. R. Essig, A. Manalaysay, J. Mardon, P. Sorensen, T. Volansky, Phys. Rev. Lett. **109**, 021301 (2012). https://doi.org/10.1103/PhysRevLett.109.021301
8. J. Angle et al., Phys. Rev. Lett. **107**, 051301 (2011). https://doi.org/10.1103/PhysRevLett.110.249901, https://doi.org/10.1103/PhysRevLett.107.051301. [Erratum: Phys. Rev. Lett. **110**, 249901 (2013)]
9. E. Aprile et al., Phys. Rev. D **94**(9), 092001 (2016). https://doi.org/10.1103/PhysRevD.94.092001, https://doi.org/10.1103/PhysRevD.95.059901. [Erratum: Phys. Rev. D **95**(5), 059901 (2017)]
10. P. Agnes, et al., Constraints on sub-GeV dark-matter–electron scattering from the darkSide-50 experiment. Phys. Rev. Lett. **121**(11), 111303 (2018). https://doi.org/10.1103/PhysRevLett.121.111303
11. A. Bernstein, R. Essig, M. Fernandez-Serra, A. Kopec, R. Lang, J. Long, K. Ni, P. Sorensen, J. Xu, LBECA: Low background electron counting apparatus (Unpublished)
12. J. Tiffenberg, M. Sofo-Haro, A. Drlica-Wagner, R. Essig, Y. Guardincerri, S. Holland, T. Volansky, T.T. Yu, Phys. Rev. Lett. **119**(13), 131802 (2017). https://doi.org/10.1103/PhysRevLett.119.131802
13. M. Crisler, R. Essig, J. Estrada, G. Fernandez, J. Tiffenberg, M. Sofo haro, T. Volansky, T.T. Yu, SENSEI: First direct-detection constraints on sub-GeV dark matter from a surface run. Phys. Rev. Lett. **121**(6), 061803 (2018). https://doi.org/10.1103/PhysRevLett.121.061803
14. R.K. Romani et al., Appl. Phys. Lett. **112**, 043501 (2018). https://doi.org/10.1063/1.5010699

15. R. Agnese et al., First dark matter constraints from a super CDMS single-charge sensitive detector. Phys. Rev. Lett. **121**(5), 051301 (2018). https://doi.org/10.1103/PhysRevLett.122. 069901, https://doi.org/10.1103/PhysRevLett.121.051301. [Erratum: Phys. Rev. Lett. **122**(6), 069901 (2019)]
16. R. Essig, M. Sholapurkar, T.T. Yu, Phys. Rev. D **97**(9), 095029 (2018). https://doi.org/10.1103/ PhysRevD.97.095029
17. R. Essig, T. Volansky, T.T. Yu, Phys. Rev. D **96**(4), 043017 (2017). https://doi.org/10.1103/ PhysRevD.96.043017
18. J. Billard, L. Strigari, E. Figueroa-Feliciano, Phys. Rev. D **89**(2), 023524 (2014). https://doi. org/10.1103/PhysRevD.89.023524
19. J.H. Chang, R. Essig, S.D. McDermott, Supernova 1987A constraints on sub-GeV dark sectors, millicharged particles, the QCD axion, and an axion-like particle. JHEP **09**, 051 (2018). https:// doi.org/10.1007/JHEP09(2018)051
20. G.G. Raffelt, *Stars as Laboratories for Fundamental Physics* (1996). http://wwwth.mpp.mpg. de/members/raffelt/mypapers/199613.pdf
21. R. Essig, M. Fernandez-Serra, J. Mardon, A. Soto, T. Volansky, T.T. Yu, JHEP **05**, 046 (2016). https://doi.org/10.1007/JHEP05(2016)046
22. H. Vogel, J. Redondo, JCAP **1402**, 029 (2014). https://doi.org/10.1088/1475-7516/2014/02/ 029
23. A.A. Prinz et al., Phys. Rev. Lett. **81**, 1175 (1998). https://doi.org/10.1103/PhysRevLett.81. 1175
24. S.D. McDermott, H.B. Yu, K.M. Zurek, Phys. Rev. D **83**, 063509 (2011). https://doi.org/10. 1103/PhysRevD.83.063509
25. S. Davidson, S. Hannestad, G. Raffelt, JHEP **05**, 003 (2000). https://doi.org/10.1088/1126-6708/2000/05/003
26. T. Emken, R. Essig, C. Kouvaris, M. Sholapurkar, Direct detection of strongly interacting sub-GeV dark matter via electron recoils. JCAP **1909**(09), 070 (2019). https://doi.org/10.1088/ 1475-7516/2019/09/070
27. T. Emken, C. Kouvaris, I.M. Shoemaker, Phys. Rev. D **96**(1), 015018 (2017). https://doi.org/ 10.1103/PhysRevD.96.015018
28. T. Emken, C. Kouvaris, JCAP **1710**(10), 031 (2017). https://doi.org/10.1088/1475-7516/2017/ 10/031
29. B.J. Kavanagh, R. Catena, C. Kouvaris, JCAP **1701**(01), 012 (2017). https://doi.org/10.1088/ 1475-7516/2017/01/012
30. J.D. Bowman, A.E.E. Rogers, R.A. Monsalve, T.J. Mozdzen, N. Mahesh, Nature **555**(7694), 67 (2018). https://doi.org/10.1038/nature25792
31. R. Barkana, N.J. Outmezguine, D. Redigolo, T. Volansky, Strong constraints on light dark matter interpretation of the EDGES signal. Phys. Rev. **D98**(10), 103005 (2018). https://doi. org/10.1103/PhysRevD.98.103005
32. A. Falkowski, K. Petraki, *21cm Absorption Signal From Charge Sequestration* (2018)

FASER and the Search for Light and Weakly Interacting Particles

Jonathan L. Feng

Abstract For decades, the leading examples of new physics targets at particle colliders were particles with TeV-scale masses and $\mathcal{O}(1)$ couplings to the standard model. More recently, however, there is a growing and complementary interest in new particles that are much lighter and more weakly coupled. I review the motivations for this shift and the importance of renormalizable portals. I then present FASER, a proposed LHC experiment that is specifically designed to discover light and weakly interacting particles, including those that interact through renormalizable portal interactions.

Keywords Dark matter · Long-lived particles · LHC · FASER

1 Introduction

Since the 1930s, beginning with the work of Ernest Lawrence and others, particle accelerators have been the workhorse tool for discovering new particles. With each significant increase in collision energy, new particles have been produced, providing profound insights into the fundamental building blocks of the universe. In the last few decades, as colliders have approached and reached TeV energies, the expectation for new particles has again been strong, with most of the attention focused on heavy particles with TeV-scale masses and $\mathcal{O}(1)$ couplings to the standard model.

More recently, however, there is a growing interest in new particles that are much lighter and more weakly coupled [1]. This complementary direction has many motivations. From the point of view of astrophysics and cosmology, the WIMP miracle continues to provide a compelling reason to search for WIMP dark matter with TeV masses. However, since the thermal relic density scales as m^2/g^4, where m and g are the dark matter's mass and coupling strength, respectively, light and weakly interacting particles may also yield the correct thermal relic density [2, 3]. The existence of

J. L. Feng (✉)
Department of Physics and Astronomy, University of California, Irvine, CA 92697-4575, USA
e-mail: jlf@uci.edu
URL: http://hep.ps.uci.edu/~jlf/

© Springer Nature Switzerland AG 2019
R. Essig et al. (eds.), *Illuminating Dark Matter*, Astrophysics and Space Science Proceedings 56, https://doi.org/10.1007/978-3-030-31593-1_8

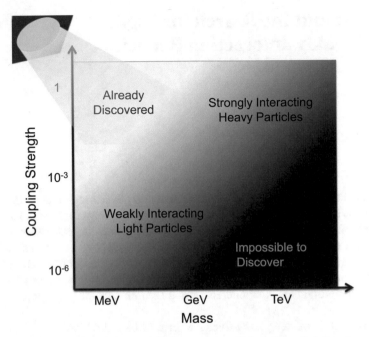

Fig. 1 The lamppost landscape for particle discovery at colliders. Strongly interacting and light particles have already been discovered; weakly interacting and heavy particles are beyond reach. The frontier for new particle discovery, therefore, lies along the diagonal and includes the traditional target of strongly interacting and heavy particles and the new target of weakly interacting and light particles

dark matter and the appeal of thermal relics with the observed abundance therefore also favors searching for light and weakly interacting particles.

From the viewpoint of particle physics, there are also strong motivations for searches for this new class of particle. Searches for new TeV-scale particles at the LHC and elsewhere have come up empty so far. These searches remain of great interest, especially given upcoming runs of the LHC and HL-LHC, but at the same time, it is natural to look elsewhere. Light and weakly interacting particles are of interest in part because they may resolve outstanding discrepancies between theory and low-energy experiments [4–6]. But perhaps most important, light and weakly interacting particles are amenable to experimental searches; see Fig. 1. As evident from the contributions of Bertrand Echenard, Mauro Raggi, and others to these proceedings, this possibility has opened the floodgates to innovative ideas for accelerator experiments that are relatively small, cheap, and fast, and may nevertheless have revolutionary implications for particle physics and cosmology.

2 Renormalizable Portals

Perhaps, the most natural origin for light and weakly interacting particles is a dark sector, containing dark matter and also possibly other matter and forces, and interacting only gravitationally with the standard model at tree level.

At loop-level, mediator particles with both standard model and dark sector interactions may induce dark sector-standard model interactions. What sort of interactions are most likely? For most such interactions, the induced interaction decouples as the mediator particle becomes heavy. However, for renormalizable interactions, this is not the case. There are, in fact, only a few possible renormalizable interactions:

- Spin 1 dark gauge bosons may interact through the kinetic mixing term $F_{\mu\nu}F_D^{\mu\nu}$. These interactions imply the existence of dark photons [7–9] with couplings to standard model fermions proportional to $q_f\epsilon$, where q_f is the fermion's charge, and ϵ is a small kinetic mixing parameter.
- Spin 0 dark scalars may interact through the quartic scalar coupling $h^\dagger h \phi_D^\dagger \phi_D$ [10]. These interactions imply the existence of dark Higgs bosons with couplings to standard model fermions proportional to $m_f \sin\theta$, where m_f is the fermion's mass, and $\sin\theta$ is a small mixing angle.
- Spin 1/2 dark fermions may interact through the Yukawa coupling $hL\psi_D$. These interactions imply the existence of dark fermions, also known as sterile neutrinos [11] or heavy neutral leptons, which mix with standard model neutrinos with a small mixing angle $\sin\theta_\nu$.

The importance of these renormalizable portal interactions is that they are non-decoupling and so may be significant even if the mediator particles have GUT- or Planck-scale masses. They rely only on the fact that the mediators exist, not that they be light. Such interactions are therefore generic in this sense and provide an organizing principle that focuses attention on a small number of dark sector candidates.

3 FASER

The possibility of light and weakly interacting new particles has motivated a number of new initiatives at particle colliders. Here, we focus on FASER [12–21], a proposed small and inexpensive experiment designed to search for light and weakly interacting particles at the LHC. Other LHC experiments with similar physics goals include the existing experiments LHCb [22, 23] and NA62 [24], as well as the proposed experiments SHiP [25], MATHUSLA [26], and CODEX-b [27], and there are, of course, also exciting opportunities at other laboratories around the world.

In contrast to heavy and strongly interacting particles, light and weakly interacting particles are dominantly produced along the beam collision axis and are typically long-lived particles (LLPs), traveling hundreds of meters before decaying. To exploit

Fig. 2 FASER's Location. Left: FASER's location is indicated by the red star in service tunnel TI12, 480 m east of the ATLAS interaction point. Credit: CERN Geographical Information System. Right: The view, looking toward the west, of FASER as it will be installed in tunnel TI12. The floor shown will be lowered by 45 cm to allow FASER to be centered on the beam collision axis. Credit: CERN Site Management and Buildings Department

both of these properties, FASER is to be located along the beam collision axis, 480 m downstream from the ATLAS interaction point (IP). At this location, FASER and a larger successor, FASER 2, will enhance the LHC's discovery potential by providing exceptional sensitivity to dark photons, dark Higgs bosons, heavy neutral leptons, axion-like particles, and many other proposed new particles.

3.1 Location and Timeline

FASER will be located 480 m downstream from the ATLAS IP in service tunnel TI12, as shown in Fig. 2. A similar tunnel, TI18, on the other side of ATLAS is also possible. These tunnels were formerly used to connect the SPS to the LEP tunnel, but are currently empty and unused.

The proposed timeline is for FASER to be installed during Long Shutdown 2 (LS2) from 2019 to 20, in time to collect data during Run 3 of the 14 TeV LHC from 2021 to 23. FASER's cylindrical active decay volume has a radius $R = 10$ cm and length $L = 1.5$ m, and the detector's total length is under 5 m. To allow FASER to maximally intersect the beam collision axis, the floor of TI12 should be lowered by 45 cm. This will not disrupt essential services, and no other excavation is required. FASER will run concurrently with the LHC and require no beam modifications. Its interactions with existing experiments are limited only to requiring bunch crossing timing and luminosity information from ATLAS.

If FASER is successful, a larger version, FASER 2, with a cylindrical active decay volume with radius $R = 1$ m and length $L = 5$ m, could be installed during LS3 from 2024 to 25 and take data in the 14 TeV HL-LHC era, starting in 2026. FASER 2 would require extending TI12 or TI18 or widening the staging area UJ18.

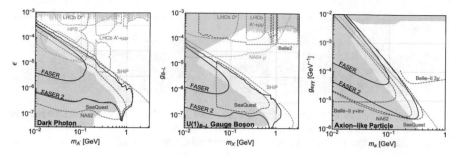

Fig. 3 Sensitivity reaches for FASER (Run 3) and FASER 2 (HL-LHC) for dark photons (left), $U(1)_{B-L}$ gauge bosons (center), and axion-like particles (right). The gray-shaded regions are excluded by current bounds, and the projected reaches of other experiments are also shown

3.2 Signals and Discovery Potential

The FASER signal is LLPs that are produced at or close to the IP, travel along the beam collision axis, and decay visibly in FASER:

$$pp \rightarrow \text{LLP} + X, \quad \text{LLP travels} \sim 480 \text{ m}, \quad \text{LLP} \rightarrow e^+e^-, \mu^+\mu^-, \gamma\gamma \ldots \quad (1)$$

These signals are striking: two oppositely charged tracks (or two photons) with $\sim$ TeV energies that start inside the detector and have a combined momentum that points back through 100 m of concrete and 90 m of rock to the IP.

The sensitivity reach of FASER has been investigated for a large number of new physics scenarios. Examples are shown in Fig. 3. FASER will have the potential to discover a broad array of new particles, including dark photons, other light gauge bosons, and axion-like particles. FASER 2 will extend FASER's physics reach in these models to larger masses and also probe currently uncharted territory for dark Higgs bosons, heavy neutral leptons, and many other possibilities.

3.3 Detector and Backgrounds

The FASER signals are two extremely energetic ($\sim$ TeV) coincident tracks or photons that start inside the detector and point back to the ATLAS IP. Muons and neutrinos are the only known particles that can transport such energies through the 190 m of concrete and rock between the IP and FASER. Muons entering the detector can be vetoed, and preliminary estimates show that muon-associated radiative processes may be reduced to negligible levels. Neutrinos may interact in the detector, but given the requirement of TeV energies and small neutrino interactions, neutrino-induced backgrounds are also negligible. The layout of the FASER detector is illustrated in Fig. 4.

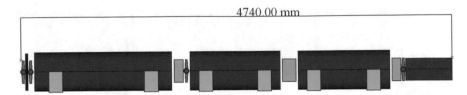

Fig. 4 Layout of the FASER detector. Particles produced at the ATLAS IIP enter from the left. The detector components include scintillators for vetoing and triggering (gray), 0.5 T dipole magnets (red), tracking stations (blue), and an electromagnetic calorimeter (purple)

Recently a FLUKA study [28–30] from the CERN Sources, Targets, and Interactions group has been carried out to assess possible backgrounds and the radiation level in the FASER location. The study shows that no high energy (>100 GeV) particles are expected to enter FASER from proton showers in the dispersion suppressor or from beam–gas interactions. In addition, the radiation level expected at the FASER location is very low due to the dispersion function in the LHC cell closest to FASER.

An emulsion detector and a battery-operated radiation monitor were installed at the FASER site in June 2018. The results from these first *in situ* measurements will complement and validate the background estimates and inform future work, which includes refining background estimates, evaluating signal efficiencies, and optimizing the detector.

Acknowledgements I am grateful to the Simons Foundation for its generous support of this Symposium and to the members of the FASER Collaboration for their many valuable contributions to this work. This work is supported in part by NSF Grant No. PHY-1620638 and in part by Simons Investigator Award #376204.

References

1. M. Battaglieri, et al., unpublished (2017)
2. C. Boehm, P. Fayet, Nucl. Phys. B **683**, 219 (2004). https://doi.org/10.1016/j.nuclphysb.2004.01.015
3. J.L. Feng, J. Kumar, Phys. Rev. Lett. **101**, 231301 (2008). https://doi.org/10.1103/PhysRevLett.101.231301
4. G.W. Bennett et al., Phys. Rev. D **73**, 072003 (2006). https://doi.org/10.1103/PhysRevD.73.072003
5. R. Pohl et al., Nature **466**, 213 (2010). https://doi.org/10.1038/nature09250
6. A.J. Krasznahorkay et al., Phys. Rev. Lett. **116**(4), 042501 (2016). https://doi.org/10.1103/PhysRevLett.116.042501
7. L.B. Okun, Sov. Phys. JETP **56**, 502 (1982). [Zh. Eksp. Teor. Fiz. 83, 892 (1982)]
8. P. Galison, A. Manohar, Phys. Lett. **136B**, 279 (1984). https://doi.org/10.1016/0370-2693(84)91161-4
9. B. Holdom, Phys. Lett. B **166**, 196 (1986). https://doi.org/10.1016/0370-2693(86)91377-8
10. B. Patt, F. Wilczek (2006)
11. H. Fritzsch, M. Gell-Mann, P. Minkowski, Phys. Lett. **59B**, 256 (1975). https://doi.org/10.1016/0370-2693(75)90040-4

12. J.L. Feng, I. Galon, F. Kling, S. Trojanowski, Phys. Rev. D **97**(3), 035001 (2018). https://doi.org/10.1103/PhysRevD.97.035001
13. J.L. Feng, I. Galon, F. Kling, S. Trojanowski, Phys. Rev. D **97**(5), 055034 (2018). https://doi.org/10.1103/PhysRevD.97.055034
14. B. Batell, A. Freitas, A. Ismail, D. Mckeen, Phys. Rev. D **98**(5), 055026 (2018). https://doi.org/10.1103/PhysRevD.98.055026
15. F. Kling, S. Trojanowski, Phys. Rev. D **97**(9), 095016 (2018). https://doi.org/10.1103/PhysRevD.97.095016
16. J.C. Helo, M. Hirsch, Z.S. Wang, JHEP **07**, 056 (2018). https://doi.org/10.1007/JHEP07(2018)056
17. M. Bauer, P. Foldenauer, J. Jaeckel, JHEP **07**, 094 (2018). https://doi.org/10.1007/JHEP07(2018)094
18. J.L. Feng, I. Galon, F. Kling, S. Trojanowski, Phys. Rev. D **98**(5), 055021 (2018). https://doi.org/10.1103/PhysRevD.98.055021
19. A. Ariga, T. Ariga, J.T. Boyd, D.W. Casper, J.L. Feng, I. Galon, S. Hsu, F. Kling, H. Otono, B. Petersen, O. Sato, A.M. Soffa, J.R. Swaney, S. Trojanowski, Letter of intent: FASER - forward search experiment at the LHC. Technical report. CERN-LHCC-2018-030. LHCC-I-032, CERN, Geneva (2018). https://cds.cern.ch/record/2642351
20. A. Berlin, F. Kling, Phys. Rev. D **99**(1), 015021 (2019). https://doi.org/10.1103/PhysRevD.99.015021
21. D. Dercks, J. de Vries, H.K. Dreiner, Z.S. Wang, Phys. Rev. D **99**(5), 055039 (2019). https://doi.org/10.1103/PhysRevD.99.055039
22. P. Ilten, J. Thaler, M. Williams, W. Xue, Phys. Rev. D **92**(11), 115017 (2015). https://doi.org/10.1103/PhysRevD.92.115017
23. P. Ilten, Y. Soreq, J. Thaler, M. Williams, W. Xue, Phys. Rev. Lett. **116**(25), 251803 (2016). https://doi.org/10.1103/PhysRevLett.116.251803
24. E. Cortina Gil, et al., JINST **12**(05), P05025 (2017). https://doi.org/10.1088/1748-0221/12/05/P05025
25. S. Alekhin et al., Rep. Prog. Phys. **79**, 124201 (2016). https://doi.org/10.1088/0034-4885/79/12/124201
26. D. Curtin et al., unpublished (2018)
27. V.V. Gligorov, S. Knapen, M. Papucci, D.J. Robinson, Phys. Rev. D **97**, 015023 (2018). https://doi.org/10.1103/PhysRevD.97.015023
28. A. Ferrari, P.R. Sala, A. Fasso, J. Ranft, unpublished (2005)
29. T.T. Böhlen, F. Cerutti, M.P.W. Chin, A. Fassa, A. Ferrari, P.G. Ortega, A. Mairani, P.R. Sala, G. Smirnov, V. Vlachoudis, Nucl. Data Sheets **120**, 211 (2014). https://doi.org/10.1016/j.nds.2014.07.049
30. M. Sabate-Gilarte, F. Cerutti, A. Tsinganis, unpublished (2018)

Interplay of Dark Matter Direct Detection and Neutrino Experiments

Roni Harnik

Abstract Dark matter detectors are approaching the neutrino floor. As standard model neutrino rates get close to being probed, dark matter experiments begin to probe neutrino physics in an interesting way, which I review here. I also present some frameworks in which deep neutrino detectors may serve as dark matter direct detection experiments.

1 Introduction

Take a look around. Dark matter is traversing all parts of the room, including the space between you and the computer screen, at a high rate. This mere fact is a driving motivation behind the effort of dark matter direct detection. A second glance brings to mind that neutrinos are also traversing the same space, and that this is driving a different effort of detecting and characterizing solar and atmospheric neutrinos.

These two experimental efforts, in which scientists search for feeble particles crossing their laboratories have some common features. Both require going deep underground and the construction of clean and sensitive detectors with large exposures and low backgrounds. There are, of course, some important differences among these detectors, including the energy threshold, which we will discuss below. In this writeup, I review some connections between dark matter searches and neutrino experiments, showing that dark matter experiments can shed light of neutrino physics and vice versa.

2 Neutrinos in Dark Matter Detectors

Neutrinos are known to play a role in direct detection, providing the irreducible background to dark matter scattering. This so-called "neutrino floor" in nuclear recoils comes from solar neutrinos for low DM masses and atmospheric neutrinos

R. Harnik (✉)
Department of Theoretical Physics, Fermilab, Batavia, IL, USA
e-mail: roni@fnal.gov

© Springer Nature Switzerland AG 2019
R. Essig et al. (eds.), *Illuminating Dark Matter*, Astrophysics and Space
Science Proceedings 56, https://doi.org/10.1007/978-3-030-31593-1_9

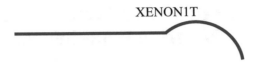

XENON1T

Fig. 1 Neutrinos can show up in dark matter experiments above the neutrino floor. Shown in red are the electron recoil spectra in several experiments taken from [2], with the recent spectrum measured by XENON1T added schematically (the precise result is presented in the contribution by Lang in this volume). The spectrum expected from standard model solar neutrinos is in solid black. The solid colorful curves (A–D) are the solar neutrino spectra for several new physics models discussed in the text

for DM above a few GeV [1]. For electron recoils, solar neutrinos dominate (see also the contribution of R. Essig in this volume).

As dark matter searches come close to the ν-floor, the interpretation of potential discoveries as coming from new ν physics becomes increasingly plausible. Turning this statement around, dark matter detectors are already capable of placing interesting limits of models of new physics in the neutrino sector.

This is apparent in the current limits on electron recoils. In Fig. 1, I show the observed electron recoil spectrum observed by several dark matter as well as neutrino experiments, taken from [2]. Data include COGENT, DAMA (unmodulated) and XENON100 spectra, as well as the data from Borexino which has observed the solar Boron-8 spectrum [3]. The contribution from SM solar neutrinos is shown as a black solid line. To update the original plot, I have added a schematic red line showing roughly the most recent measurement from XENON1T [4], which represents about two orders of magnitude improvement on the XENON100 background rate. A more precise representation of this measurement as well as a detailed description of XENON1T can be found in R. Lang's contribution to this volume. XENON1T is thus within an order of magnitude of the ν-floor in electron recoil.

From this state of affairs, it is clear that the amount of room for new physics is becoming smaller. In Fig. 1, I also show several spectra from new physics models which lead to an enhanced scattering rate at low recoils. Curve A shows the recoil spectrum in the case that the neutrino possesses a magnetic dipole moment around that which is allowed by current laboratory experiments (the limit by GEMMA [5] is about 10% lower). In this case, the differential cross section is

$$\frac{d\sigma}{dE_r} = \mu_\nu^2 \alpha \left(\frac{1}{E_r} - \frac{1}{E_\nu} \right) \tag{1}$$

where μ_ν is the neutrino dipole moment and E_r is the recoil energy of the electron. At high recoil energies, the dipole induced scattering is lower than the SM rate and in agreement with the Borexino rate. However, due to the E_r^{-1} falloff, the rate is higher at low recoil energies. From this, we can estimate that an analysis by XENON1T can potentially improve the limit in dipole moments to $\sim 1.5 \times 10^{-11}$ times a Bohr Magneton, about a factor of two better than the current limits.

One can also consider models with a faster falling spectrum. For example, curves B, C, and D of Fig. 1 are the spectra in a model with a new, very light $B - L$ gauge boson which is mediating a new interaction between neutrinos and electrons. The cross section is

$$\frac{d\sigma}{dE_r} = \frac{g_{B-L}^4 m_e}{4\pi(2m_e^2 E_r^2 + m_{B-L}^2)^2},\qquad(2)$$

where m_{B-L} and g_{B-L} are the mass and coupling, we have dropped subleading terms in E_r/E_ν as well as interference with the SM process which is unimportant at most recoil energies. If the mass of the gauge boson is small, the cross section falls as E_r^{-2}. This behavior is due to the $1/(q^2 - m_{B-L}^2)$ propagator in the amplitude, with $q^2 = 2m_e E_r$. Again, we see that XENON1T can constrain some models which are allowed by Borexino. A by-eye estimate suggests that in the limit where the mass of the $B - L$ is negligible (less than about 30 keV), XENON1T can place a limit of $g_{B-L} \lesssim 2 \times 10^{-6}$. This is almost an order of magnitude improvement on Borexino's limit and about factor of two stronger than the reactor experiment GEMMA.

It is interesting to consider a scenario in which the next generation of XENON experiments lowers their electron recoil backgrounds further and uncovers an excess above the solar neutrino floor. As was pointed out during the symposium, we will immediately entertain both the possibility of dark matter and that of new ν physics. Fortunately, this can be disentangled with reactor experiments. Nuclear reactors are a brighter source of neutrinos than the sun (to those that stand within 100 m of the core). A low threshold detector near a reactor, such as GEMMA, can thus place strong limits or distinguish whether an excess is coming from dark matter or neutrinos. An exciting experiment in this class is CONNIE [6], an experiment dedicated to discovering coherent ν-nucleus scattering in reactors. CONNIE consist of a low threshold CCD detector, similar to that used in the DAMIC dark matter experiment [7], placed 30 m away from a reactor core in Brazil. The energy threshold in this detector will be about an order of magnitude lower than that of GEMMA, leading to stronger limits on the new physics models discussed above. CONNIE results are expected this year. In the future, the breakthrough in CCD technology described in the contribution of J. Tiffenberg to this volume will push the energy threshold even lower, leading to enhanced sensitivity to new physics.

3 Dark Matter in Neutrino Detectors

Dark matter direct detection requires detectors that are extremely low in the background, with large exposures, and with low energy thresholds. The later is needed because the energy available for direct detection is the kinetic energy in the DM-nucleus system, $\sim\mu v^2$, where μ is the reduced mass and v is the typical DM velocity, of order 10^{-3}. The required thresholds are thus at most in the range of 10's of keV (but less for light dark matter masses).

Table 1 An incomplete list of neutrino detector with their approximate mass and energy threshold. The top portion lists existing detectors while the bottom list planned experiments. The question marks serve to remind that the thresholds of future detectors are rough guesses

Detector	Approx. mass (ton)	Approx. Threshold
XENON1T	1	Few keV
Borexino	10^2	150 keV
SNO	10^3	MeV
SuperK	5×10^4	6 MeV
IceCube	10^7	10 GeV
SNO+	10^3	200 keV?
JUNO	few 10^2	200 keV?
DUNE	3×10^4	1 MeV?
HyperK	5×10^5	6 MeV?

Neutrino detectors, like their DM counterparts, are deep, clean, and have large exposures, but have larger threshold. In Table 1, a list of several present and future neutrino detectors is shown (including XENON1T, which, as established in the previous section, is a neutrino detector of sorts). It is interesting to note that the detector masses are much larger for the neutrino detectors. Though the energy thresholds are too high for DM detection, it is remarkable that some neutrino detectors are close to the mark.

Motivated by this set of experiments, I will briefly mention two frameworks for dark matter which can be probed by neutrino detectors. The first is accessible to the lower threshold detectors such as Borexino, JUNO, and SNO+ while the second model may be probed by all of the experiments. Both of these frameworks rely on simple models that lead to distinct signals in the experiment (in spirit with the point of view presented by K. Zurek in this volume).

3.1 Luminous Dark Matter and Daily Modulation

Models of inelastic dark matter [8] (iDM) are simple extensions of the minimal elastically scattering dark matter. As such, they remain a well-motivated framework beyond their original intent of explaining the DAMA modulation signal [9]. In particular, iDM can provide a simple explanation of why a canonical WIMP, one that is part of an EW doublet (like a TeV Higgsino), has failed to show up in direct detection experiments. Direct detection experiments cannot probe iDM if the mass splitting in the DM system, δ, is above the available kinetic energy μv^2. For heavy iDM, the limiting factor is that μ is set by the target nucleus mass.

Low threshold neutrino detectors can extend the reach of these searches to higher δ by making use of heavy elements that are in the rock [10]. Dark matter can up-scatter into the excited state on a lead nucleus. Due to the limited phase space, the decay back into the ground state can be long-lived to be somewhat displaced from the scatter site (a fraction of a second is not untypical). A detector such as Borexino can thus detect a monoenergetic photon, of energy $\sim\delta$, emitted in the decay $\chi_2 \rightarrow \chi_1\gamma$. This two-step process has been studied in other contexts and is dubbed luminous DM [11, 12]. It should be noted that so long as the lifetime of χ_2 is shorter than 10 s, the traverse time of the earth, the rate for decays will be set by the rate of excitations, and is thus parametrically similar to inelastic direct detection, but due to the lead nuclei in the surrounding rock.

There is, however, one interesting difference in the rate, which allows for an additional experimental handle in the search. For heavy dark matter, there is a strong daily modulation in the signal event rate. Because experiments are situated close to the earth's surface (as compared to the lifetime of χ_2), the distribution of target lead nuclei is anisotropic around the detector. Second, the fast component of DM, that which has enough kinetic energy to up-scatter, is also strongly anisotropic, coming mainly from the direction of Cygnus. The third important factor is that for heavy dark matter, the up-scattering will be forward, so χ_2 is emitted from the scatter site in a cone oriented toward the incoming direction of χ_1. Combining these factors leads to the conclusion that the event rate is much higher during the time of day in which Cygnus is below the horizon at the detector location. This modulation effect and the resulting reach is studied quantitatively in [10].

3.2 Self-destruction Dark Matter

In the previous example, only neutrino detectors with thresholds as low as a few hundred keV can probe luminous dark matter. Here, I present a framework in which the largest neutrino detectors, which are also those with the highest threshold can place limits on DM. To this end, dark matter must give up more than the available kinetic energy. We thus consider Self Destructing Dark Matter (SDDM) [13].

In SDDM, dark matter (or a subcomponent of it) can completely decay inside the detector, giving up its full rest mass as visible energy. Of course, limits on the dark matter lifetime prevent this from occurring in one step. The decay thus proceeds in two steps: (1) a DM particle Ψ scatters somewhere in the earth, triggering its transition to a different particle Ψ', and (2) the properties of Ψ' are different than its parent, allowing it to fully decay in the detector.

This story begs the question: why would not Ψ transition to Ψ' earlier in its life? It is interesting to note that traversing the earth is an extremely special event in the lifetime of a DM particle. Such a particle has spent a Hubble time in the galactic halo, where the number density of particles is of order one per cm^3 (for DM around a GeV). Traversing the earth, by comparison, takes only about 10 s, 16 orders of magnitude shorter. However, the number density of particles is 23 orders of magnitude higher. The probability for transition can thus be higher in the short earth crossing than in the preceding Hubble time (though note that the transition probability will be small for both in the interesting region of parameter space). The condition above may not be sufficient to prevent transitions, since over densities such as gas clouds may also play a role, but it is certainly a required condition that our terrestrial environment fulfills.

In SDDM, a framework invariably requires a non-minimal dark sector. In particular, the symmetry that allows Ψ to be cosmologically long-lived must be undone by a single scattering. In [13], we have presented several possibilities of dark sector dynamics that allow for this. The common ingredients for these models are simple Lagrangians that, like the SM, lead to bound states with several dynamical scales. Perhaps the most elegant possibility is that Ψ is a bound state of high angular momentum that is unlikely to decay due to a centrifugal barrier. In this case, rotational symmetry, conservation of angular momentum, is stabilizing Ψ. A single scattering event can violate the symmetry, transferring some angular momentum to the bound system and allowing it to decay quickly.

The SDDM framework leads to interesting signals in large detectors such as SNO, SuperK, and DUNE, including multi-lepton events with particular kinematic properties and angular dependence. A more detailed study is presented in [13].

Acknowledgements I would like to thank the Simons Foundation and the organizers for an interesting and stimulating workshop. I would also like to thank my collaborators on the projects connected to this writeup—J. Kopp, P. Machado, Y. Grossman, O. Telem, Y. Zhang, P. Fox, J. Eby, G. Kribs, as well as the CONNIE collaboration. Fermilab is operated by Fermi Research Alliance, LLC under Contract No. DE-AC02-07CH11359 with the United States Department of Energy.

References

1. J. Billard, L. Strigari, E. Figueroa-Feliciano, Phys. Rev. D **89**(2), 023524 (2014). https://doi.org/10.1103/PhysRevD.89.023524
2. R. Harnik, J. Kopp, P.A.N. Machado, JCAP **1207**, 026 (2012). https://doi.org/10.1088/1475-7516/2012/07/026
3. M. Agostini, et al. (2017)
4. E. Aprile et al., Eur. Phys. J. C **77**(12), 881 (2017). https://doi.org/10.1140/epjc/s10052-017-5326-3
5. A.G. Beda, V.B. Brudanin, V.G. Egorov, D.V. Medvedev, V.S. Pogosov, M.V. Shirchenko, A.S. Starostin, Adv. High Energy Phys. **2012**, 350150 (2012). https://doi.org/10.1155/2012/350150
6. A. Aguilar-Arevalo et al., J. Phys. Conf. Ser. **761**(1), 012057 (2016). https://doi.org/10.1088/1742-6596/761/1/012057
7. A. Aguilar-Arevalo et al., Phys. Rev. D **94**(8), 082006 (2016). https://doi.org/10.1103/PhysRevD.94.082006
8. D. Tucker-Smith, N. Weiner, Phys. Rev. D **64**, 043502 (2001). https://doi.org/10.1103/PhysRevD.64.043502
9. J. Bramante, P.J. Fox, G.D. Kribs, A. Martin, Phys. Rev. D **94**(11), 115026 (2016). https://doi.org/10.1103/PhysRevD.94.115026
10. J. Eby, P. Fox, R. Harnik, G. Kribs

11. B. Feldstein, P.W. Graham, S. Rajendran, Phys. Rev. D **82**, 075019 (2010). https://doi.org/10.1103/PhysRevD.82.075019
12. M. Pospelov, N. Weiner, I. Yavin, Phys. Rev. D **89**(5), 055008 (2014). https://doi.org/10.1103/PhysRevD.89.055008
13. Y. Grossman, R. Harnik, O. Telem, Y. Zhang (2017)

Why I Think That Dark Matter Has Large Self-interactions

Manoj Kaplinghat

Abstract We describe a self-interacting dark matter (SIDM) model that can explain the diverse rotation curves of spiral galaxies while maintaining the success of the cold dark matter model on large scales. The explanation is economical in that it only requires one parameter, the self-interaction cross section, which is common to all galaxies. The existence of this solution is demonstrated through fits to a diverse set of 135 rotation curves from the SPARC sample. Despite the apparent diversity, the model exhibits a tight correlation between the accelerations due to the dark and luminous matter. The inferred stellar mass-to-light ratios, halo masses, and halo concentrations are consistent with independent expectations from astrophysics and cosmology.

Keywords Dark matter · Rotation curves · Galaxy formation

1 A Self-interacting Dark Matter Model

Large self-interactions are the norm in the visible sector. If dark matter is part of a hidden sector, then there is no reason to expect that dark matter will not have self-interactions. Despite this, the dominant assumption in astrophysics is that the dark matter particles only interact via gravity.

If we move away from this narrow assumption, then a whole world of possibilities opens up. However, only a limited range of these possibilities are consistent with astrophysical observations [1–6]. Here, we show that within the viable region of parameter space, the addition of dark matter self-interaction can provide a compelling and economical explanation for the observed rotation curves of spiral galaxies, which have a diverse range of shapes that has been difficult to understand [7].

We consider a SIDM model that has a large and constant cross section per unit mass (σ/m) for elastic scattering [8–12], $\sigma/m > 1\,\mathrm{cm^2/g}$ at velocities below about

M. Kaplinghat (✉)
Department of Physics and Astronomy, University of California, Irvine, CA 92697-4575, USA
e-mail: mkapling@uci.edu

© Springer Nature Switzerland AG 2019
R. Essig et al. (eds.), *Illuminating Dark Matter*, Astrophysics and Space
Science Proceedings 56, https://doi.org/10.1007/978-3-030-31593-1_10

250 km/s, which is the regime relevant for galaxies. This model has a few notable features.

- The large-scale structure predictions are the same as the CDM model and hence it inherits the successes of the CDM model automatically.
- The qualitative predictions for the dark matter halo density profile are similar for σ/m in the $1-10$ cm^2/g range [13]; in this sense, the model is not tuned.
- The large σ/m results in the halo being driven quickly toward the isothermal solution, and thus the predictions tend to be insensitive to the star formation history. This implies that different implementations of the star formation feedback will result in the same final dark matter density profile.
- Diversity of density profiles in the inner regions of galaxies is built into this model because the isothermal solution is $\rho_{\rm DM} \propto \exp(-\Phi/\sigma_{v0}^2)$ and Φ is the gravitational potential of all the matter (dark and luminous). Thus, the inner dark matter halo profile has a large spread depending on the outer halo profile (which sets the velocity dispersion of dark matter σ_{v0} and the normalization of the density profile) and the stellar density profile [14].
- When the relative velocity is around 1000 km/s, we know that σ/m must be smaller to explain the inferred central density of clusters of galaxies [15]. This velocity dependence, however, will not be relevant to explain the galactic rotation curves.

2   Diversity and Uniformity of Rotation Curves

SIDM model predicts a large range of core sizes in galactic halos [16]. To see why the diversity appears, let us consider the isothermal solution $\rho_{\rm DM} = \rho_0 \exp(-(\Phi(r) - \Phi(0))/\sigma_{v0}^2)$. This solution should match on to the Navarro–Frenk–White (NFW) density profile at large radii (see below), where the self-interactions have not had much of an impact. Thus ρ_0 depends both on the distribution of baryons (through the gravitational potential Φ) and the outer halo (which is essentially the same as that in ΛCDM).

For galaxies where the baryons do not contribute significantly to the gravitational potential, the SIDM simulations show that the halo has a cored profile with the core radius r_c scaling such that $\rho_0 r_c^2 \propto \sigma_{v0}^2$. For $\sigma/m = \mathcal{O}(1{\rm cm}^2/{\rm g})$ or larger, the core size is close to the NFW scale radius r_s. When the contribution of baryons to the gravitational potential increases, then the core size shrinks. In the limit that the stars dominate the gravitational potential, the core size is set by the stellar density profile.

Is the diversity already present in the SIDM model (described in terms of the core size above) the right kind to explain the rotation curves of galaxies? We can visually see that this is so in Fig. 1, where we have plotted the SIDM fits a representative sample of rotation curves from the SPARC sample [17]. To arrive at these fits, we have fixed $\sigma/m = 3$ cm^2/g. However, any value in the $1-10$ cm^2/g will work equally well. Even larger cross sections may fit the data, but we need more work on SIDM simulations in that range to explore that possibility.

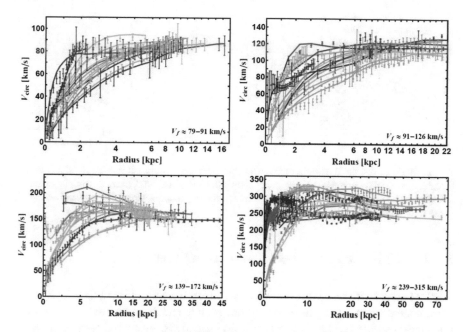

Fig. 1 SIDM fits (solid lines) over the full range of spiral galaxy masses that illustrate the diversity. The colors (blue to red) are in increasing order of surface brightness in each panel

There are four free parameters for each rotation curve fit—stellar mass-to-light ratios for the disk and bulge, two parameters describing the outer NFW halo, r_{max} and V_{max}; r_{max} is the radius where the rotation speed due to dark matter achieves its maximum value V_{max}. As alluded to before, the inner halo where interactions are rapid will be isothermal, while the outer halo where interactions have not had a significant impact will be close to NFW profile.

Matching the mass and density of the isothermal and NFW profiles at a radius r_1, where the dark matter particles have had one interaction over the age of the halo, provides a good description of the SIDM halo profiles measured in simulations. Matching the mass smoothly at r_1 with these two conditions allows ρ_0 and σ_{v0} to be related to the NFW halo parameters V_{max} and r_{max}. We note that this is an empirical recipe to smoothly join two mass profiles and neither the mass nor the density of the two profiles may be equal at r_1. The recipe works well at describing simulated halos with and without baryons.

The SIDM predictions have a regularity that is not evident in the rotation curves. However, this becomes evident when the total acceleration (g_{tot}) is plotted against the acceleration due to baryons (g_{bar}), as shown in the left panel of Fig. 2. The solid line going through the middle of the predictions is a good phenomenological description of MOND $g_{tot}/g_\dagger = x/(1 - \exp(-\sqrt{x}))$, where $x = g_{bar}/g_\dagger$ and $g_\dagger$ is the same for all galaxies [18]. The distribution of the required stellar disk mass-to-light ratios is similar in the two models (middle panel). However, the SIDM fits are generally superior (right panel).

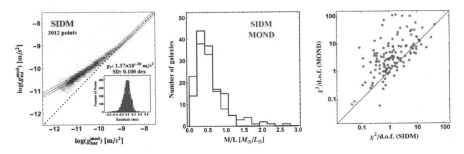

Fig. 2 SIDM model fits illustrating the uniformity (left panel) in g_{tot} (total acceleration) vs g_{bar} (acceleration due to baryons). Middle panel shows the values of the disk mass-to-light ratios for SIDM and MOND fits. The χ^2 per degree of freedom comparison between MOND and SIDM fits in the right panel

3 The Case for Large Self-interaction Strength

A SIDM model with a large cross section is a better fit to the data than CDM, given our current understanding of galaxy formation. With the addition of just one parameter, this model can fit the rotation curves of spiral galaxies over the full range of masses. It is an economical solution to a problem that has been around for more than two decades.

The diversity in the model arises from both the properties of the stellar disk and those of the outer NFW halo. The concentration–mass relation of the outer NFW halo is consistent with the expectations for the Planck cosmology [19]. Remarkably, the stellar disk mass-to-light ratios required to fit the rotation curves are distributed around 0.5 $M_\odot/L_\odot$ (in the 3.6 μm), which is consistent with the expectations of stellar population synthesis models. In addition to these features, the inferred stellar and halo masses are in excellent agreement with abundance matching results. There is also a tight correlation between the mass of the baryons (gas and stars) and the flat part of the rotation curve V_f.

The SIDM solution correctly ascribes the lowest dark matter densities to galaxies with the smallest stellar surface brightnesses and it also predicts the correct scaling of core sizes and densities with halo mass. ΛCDM models with large feedback can create dwarf and low-surface brightness galaxies with large cores [20] but they fail to simultaneously explain the presence of high-surface brightness galaxies. This could be due to the fact that if strong feedback reduces dark matter densities, then it will do the same to the (collisionless) stars.

The uniformity inherent in the SIDM solution can be seen in the tight correlation evident when the total acceleration is plotted against the acceleration due to baryons. Despite the fact that the SIDM model is not equivalent to a modification of gravity, it leads to a tight radial acceleration relation. The key factors that determine relation are the acceleration scale in ΛCDM/ΛSIDM models [21, 22] and the correlation between the luminous and dark matter due to thermalization of dark matter.

The simplicity of the solution described above, its consistency with other aspects of cosmology and galaxy formation physics, and the gain in descriptive power with the addition of just one parameter, all taken together argue for a large self-interaction cross section for dark matter particles.

References

1. J.L. Feng, M. Kaplinghat, H. Tu, H.B. Yu, JCAP **0907**, 004 (2009). https://doi.org/10.1088/1475-7516/2009/07/004
2. M.R. Buckley, P.J. Fox, Phys. Rev. D **81**, 083522 (2010). https://doi.org/10.1103/PhysRevD.81.083522
3. A. Loeb, N. Weiner, Phys. Rev. Lett. **106**, 171302 (2011). https://doi.org/10.1103/PhysRevLett.106.171302
4. A.H.G. Peter, M. Rocha, J.S. Bullock, M. Kaplinghat, Mon. Not. Roy. Astron. Soc. **430**, 105 (2013). https://doi.org/10.1093/mnras/sts535
5. J. Zavala, M. Vogelsberger, M.G. Walker, Mon. Not. Roy. Astron. Soc. Lett. **431**, L20 (2013). https://doi.org/10.1093/mnrasl/sls053
6. S. Tulin, H.B. Yu, K.M. Zurek, Phys. Rev. Lett. **110**(11), 111301 (2013). https://doi.org/10.1103/PhysRevLett.110.111301
7. K.A. Oman et al., Mon. Not. Roy. Astron. Soc. **452**(4), 3650 (2015). https://doi.org/10.1093/mnras/stv1504
8. D.N. Spergel, P.J. Steinhardt, Phys. Rev. Lett. **84**, 3760 (2000). https://doi.org/10.1103/PhysRevLett.84.3760
9. C. Firmani, E. D'Onghia, V. Avila-Reese, G. Chincarini, X. Hernandez, Mon. Not. Roy. Astron. Soc. **315**, L29 (2000). https://doi.org/10.1046/j.1365-8711.2000.03555.x
10. M. Vogelsberger, J. Zavala, A. Loeb, Mon. Not. Roy. Astron. Soc. **423**, 3740 (2012)
11. M. Rocha, A.H.G. Peter, J.S. Bullock, M. Kaplinghat, S. Garrison-Kimmel, J. Onorbe, L.A. Moustakas, Mon. Not. Roy. Astron. Soc. **430**, 81 (2013). https://doi.org/10.1093/mnras/sts514
12. S. Tulin, H.B. Yu, Phys. Rep. **730**, 1 (2018). https://doi.org/10.1016/j.physrep.2017.11.004
13. O.D. Elbert, J.S. Bullock, S. Garrison-Kimmel, M. Rocha, J. Oñorbe, A.H.G. Peter, Mon. Not. Roy. Astron. Soc. **453**(1), 29 (2015). https://doi.org/10.1093/mnras/stv1470
14. M. Kaplinghat, R.E. Keeley, T. Linden, H.B. Yu, Phys. Rev. Lett. **113**, 021302 (2014). https://doi.org/10.1103/PhysRevLett.113.021302
15. M. Kaplinghat, S. Tulin, H.B. Yu, Phys. Rev. Lett. **116**(4), 041302 (2016). https://doi.org/10.1103/PhysRevLett.116.041302
16. A. Kamada, M. Kaplinghat, A.B. Pace, H.B. Yu, Phys. Rev. Lett. **119**(11), 111102 (2017). https://doi.org/10.1103/PhysRevLett.119.111102
17. F. Lelli, S.S. McGaugh, J.M. Schombert, Astron. J. **152**, 157 (2016). https://doi.org/10.3847/0004-6256/152/6/157
18. S. McGaugh, F. Lelli, J. Schombert, Phys. Rev. Lett. **117**(20), 201101 (2016). https://doi.org/10.1103/PhysRevLett.117.201101
19. A.A. Dutton, A.V. Maccio, Mon. Not. Roy. Astron. Soc. **441**(4), 3359 (2014). https://doi.org/10.1093/mnras/stu742
20. F. Governato, A. Zolotov, A. Pontzen, C. Christensen, S.H. Oh, A.M. Brooks, T. Quinn, S. Shen, J. Wadsley, Mon. Not. Roy. Astron. Soc. **422**, 1231 (2012). https://doi.org/10.1111/j.1365-2966.2012.20696.x
21. F.C. van den Bosch, J.J. Dalcanton, Astrophys. J. **534**, 146 (2000). https://doi.org/10.1086/308750
22. M. Kaplinghat, M.S. Turner, Astrophys. J. **569**, L19 (2002). https://doi.org/10.1086/340578

Primordial Black Holes as Dark Matter: New Formation Scenarios and Astrophysical Effects

Alexander Kusenko

Abstract Scalar field instability can lead to a short matter dominated era, during which the matter is represented by large lumps of the scalar field, whose distribution exhibits large fluctuations, leading to copious production of primordial black holes (PBH). The PBH abundance can be sufficient to explain up to 100% of dark matter without violating observational constraints. Small PBH can destabilize neutron stars and contribute to r-process nucleosynthesis.

1 Introduction

Primordial black holes can account for all or part of dark matter in the early universe [1–6], and they can also seed supermassive black holes observed in centers of galaxies [7–10]. Furthermore, they could be responsible for some of the gravitational wave signals observed by LIGO [11–14].

The high-density environment in the early universe suggests that black holes may be produced if there is a sufficient degree of inhomogeneity [1–3]. However, the density perturbations that seeded the observed structures were too small for PBH formation. Some additional power could be generated on certain scales by inflaton dynamics [4], and many models have focused on this possibility [15, 16].

However, the presence of even a single scalar field (such as the Higgs field, if it has the right potential at large VEV, or some other fields, such as those predicted by supersymmetry) can result in large inhomogeneities on some scales. The origin of such inhomogeneities is in instability that causes fragmentation of a scalar condensate [17]. The instability leads to matter like state, in which the matter component is composed of large-mass lumps of the scalar field. Since the energy density in the matter component scales slower than the radiation matter density, the lumpy scalar

A. Kusenko (✉)
Department of Physics and Astronomy, University of California,
Los Angeles, CA 90095-1547, USA

Kavli Institute for the Physics and Mathematics of the Universe (WPI),
UTIAS The University of Tokyo, Kashiwa, Chiba 277-8583, Japan
e-mail: kusenko@ucla.edu

© Springer Nature Switzerland AG 2019
R. Essig et al. (eds.), *Illuminating Dark Matter*, Astrophysics and Space
Science Proceedings 56, https://doi.org/10.1007/978-3-030-31593-1_11

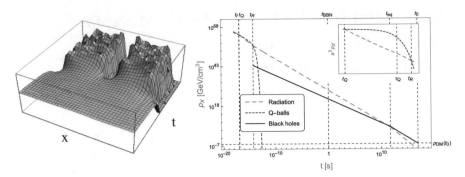

Fig. 1 Fragmentation of the scalar field (left panel, see Ref. [17]) can lead to a matter dominated stage with relatively few giant particles, which thus exhibit large density fluctuations [22, 23]. These density fluctuations, which lead to PBH production, are different from primordial density fluctuations seeding cosmic structures

field can come to dominate the energy density. The field lumps are large and relatively few, and the density fluctuations are much larger than in the case of matter made up of a huge number of small particles. Therefore, it is much more likely to find some patches of space in which the density contrast is of order one, which is necessary condition for PBH formation. Another condition, of near spherical symmetry, is also satisfied in some small subset of the universe.

2 Scalar Field Instability and PBH Formation

During inflation, scalar fields with masses smaller than the Hubble parameter develop large expectation values [18–21]. After inflation is over, the field relaxes to the minimum of its effective potential. There is a well-known instability that can set in during the coherent motion of the scalar field [17]. If the second derivative of the potential is sufficiently small or negative, an initially homogeneous condensate fragments into lumps of scalar field or Q-balls [21], as shown in the left panel of Fig. 1. The right panel shows the timeline of one such model [22], in which the scalar lumps come to dominate the energy density at time t_Q.

Eventually, the scalar lumps decay, and the radiation dominated era resumes. However, during the intermediate matter dominated era, PBH can be produced.

3 PBH Formation During a Lump-Dominated Epoch

When the "matter" is composed of relatively few giant "particles" (scalar lumps), the density fluctuations can be large. The regions of high density can give rise to black holes. This mechanism is very different from models that rely on primordial density fluctuations generated during inflation [1–6, 15]. It is also different from a model

Fig. 2 PBH mass function
in a model of Ref. [22]. See
Refs. [22, 23, 26] for
discussion of constraints and
mass functions in other
models. Solid green line
indicates the parameter space
where neutron star
disruptions by PBH can
produce up to 100% of
r-process material needed to
explain heavy element
abundances [27]

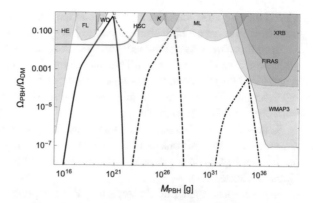

based on inhomogeneous baryogenesis [24], in which the scalar dynamics lead to formation of high-baryon-number bubbles, which collapse to black holes.

The presence of a sufficient density contrast is not yet sufficient for a black hole formation. The mass distribution in the overdense region should be spherically symmetric to a high degree [25]. The PBHs form from a small subset of the overdense regions (which, in turn, are a small subset of the total). Even though the PBH-forming configurations are rare, there is a sufficient number of them to account for all dark matter [22, 23].

The mass function of PBH produced from scalar instability is shown in Fig. 2. The PBH abundance can account for all dark matter in the mass window of 10^{20-23} g, where there are no strong constraints on the abundance of PBHs. There can also be black holes with 1–10 solar masses, which can contribute to the gravitational waves observed by LIGO. A similar scenario exists for the inflaton field, which can fragment into oscillons [26].

4   Neutron Star Genocide and Other Astrophysical Effects of PBH

Neutron stars can capture PBH, in which case the neutron star is destroyed eventually by a black hole eating it from the inside [28]. The last stages of the neutron star demise can be accompanied by a massing release of cold nuclear matter [27], which can contribute to the r-process nucleosynthesis. Rapid-capture (r-process) nucleosynthesis is needed to explain the observed abundances of heavy elements, including gold, platinum, and uranium. However, the site of r-process remains unknown, while neutron star collisions can release some neutron-rich matter, other sources may contribute to r-process. In the part of the parameter space shown in Fig. 2, neutron star genocide by PBH can account for up to 100% of the needed r-process.

PBH contribution to r-process nucleosynthesis is consistent with the observed distribution of heavy elements in dwarf spheroidal galaxies [29]: since the capture of a PBH on a neutron star is a rare event, one expects that roughly one in ten ultrafaint dwarfs should have a high abundance of heavy elements [27].

In addition to r-process nucleosynthesis, the presence of PBH can result in several additional astrophysical effects. The last stages of neutron star destructions cause the magnetic field of the star to undergo a transformation on the time scales of a few milliseconds. This results in a radio pulse whose duration and energy are consistent with observed fast radio bursts.

Released nuclear matter, heated by beat decays, reaches temperatures at which some fraction of positrons can be produced. These low-energy positrons eventually annihilate, and their population can explain the observed 511 keV line from the Galactic Center [27].

Regardless of their initial size, small PBH captured on neutron stars transform into black holes with masses from 1 to 2 $M_\odot$ [27, 30]. Since astrophysical black holes are expected to have larger masses, detection of a population of black holes with masses $(1-2)M_\odot$ would imply the existence of PBH.

5   All Dark Matter in the Form of PBH?

There is an open mass window, shown in Fig. 2, in which all dark matter can be made up by primordial black holes. Several techniques used to rule out PBH at masses below $10^{-13}M_\odot$ and above $10^{-10}M_\odot$ are ineffective in this mass window. For example, optical microlensing does not work for black holes whose event horizons are smaller than the wavelength of light [31, 32]. PBH in this mass window can be produced in the early universe in a number of models that make few assumptions beyond inflation and, possibly, an additional scalar field [22, 23, 26].

6   Conclusion

A new class of models for PBH formation, based on the scalar field instability in the early universe, makes PBH formation a natural and fairly generic phenomenon. There is a scalar field in the Standard Model, namely, the Higgs field, and theories beyond the Standard Model typically predict a large number of scalar fields, for example, from supersymmetry.

Acknowledgements I thank the Simons Foundation for support and hospitality of the Simons Symposium, which stimulated many new ideas, including a new project that K. Abazajian and I have started at Schloss Elmau. This work was supported by the U.S. Department of Energy Grant No. DE-SC0009937 as well as World Premier International (WPI) Initiative, MEXT, Japan.

References

1. Y.B. Zel'dovich, I.D. Novikov, Sov. Astron. **10**, 602 (1967)
2. S. Hawking, Mon. Not. Roy. Astron. Soc. **152**, 75 (1971)
3. B.J. Carr, S.W. Hawking, Mon. Not. Roy. Astron. Soc. **168**, 399 (1974)
4. J. Garcia-Bellido, A.D. Linde, D. Wands, Phys. Rev. D **54**, 6040 (1996). https://doi.org/10.1103/PhysRevD.54.6040
5. P.H. Frampton, M. Kawasaki, F. Takahashi, T.T. Yanagida, JCAP **1004**, 023 (2010). https://doi.org/10.1088/1475-7516/2010/04/023
6. K. Inomata, M. Kawasaki, K. Mukaida, Y. Tada, T.T. Yanagida (2016)
7. R. Bean, J. Magueijo, Phys. Rev. D **66**, 063505 (2002). https://doi.org/10.1103/PhysRevD.66.063505
8. M. Kawasaki, A. Kusenko, T.T. Yanagida, Phys. Lett. B **711**, 1 (2012). https://doi.org/10.1016/j.physletb.2012.03.056
9. S. Clesse, J. García-Bellido, Phys. Rev. D **92**(2), 023524 (2015). https://doi.org/10.1103/PhysRevD.92.023524
10. A.D. Dolgov, Usp. Fiz. Nauk **188**(2), 121 (2018).https://doi.org/10.3367/UFNe.2017.06.038153. [Phys. Usp.61,no.2,115(2018)]
11. S. Clesse, J. García-Bellido, Phys. Dark Univ. **15**, 142 (2017). https://doi.org/10.1016/j.dark.2016.10.002
12. S. Bird, I. Cholis, J.B. Muñoz, Y. Ali-Haïmoud, M. Kamionkowski, E.D. Kovetz, A. Raccanelli, A.G. Riess, Phys. Rev. Lett. **116**(20), 201301 (2016). https://doi.org/10.1103/PhysRevLett.116.201301
13. M. Sasaki, T. Suyama, T. Tanaka, S. Yokoyama, Phys. Rev. Lett.**117**(6), 061101 (2016).https://doi.org/10.1103/PhysRevLett.121.059901, https://doi.org/10.1103/PhysRevLett.117.061101. [Erratum: Phys. Rev. Lett.121,no.5,059901(2018)]
14. J. Georg, S. Watson, JHEP **09**, 138 (2017). https://doi.org/10.1007/JHEP09(2017)138. [JHEP09,138(2017)]
15. MYu. Khlopov, Res. Astron. Astrophys. **10**, 495 (2010). https://doi.org/10.1088/1674-4527/10/6/001
16. B. Carr, T. Tenkanen, V. Vaskonen, Phys. Rev. D **96**(6), 063507 (2017). https://doi.org/10.1103/PhysRevD.96.063507
17. A. Kusenko, M.E. Shaposhnikov, Phys. Lett. B **418**, 46 (1998). https://doi.org/10.1016/S0370-2693(97)01375-0
18. T.S. Bunch, P.C.W. Davies, Proc. Roy. Soc. Lond. **A360**, 117 (1978). https://doi.org/10.1098/rspa.1978.0060
19. A.D. Linde, Phys. Lett. B **116**, 335 (1982). https://doi.org/10.1016/0370-2693(82)90293-3
20. A.A. Starobinsky, J. Yokoyama, Phys. Rev. D **50**, 6357 (1994). https://doi.org/10.1103/PhysRevD.50.6357
21. M. Dine, A. Kusenko, Rev. Mod. Phys. **76**, 1 (2003). https://doi.org/10.1103/RevModPhys.76.1
22. E. Cotner, A. Kusenko (2016)
23. E. Cotner, A. Kusenko, Phys. Rev. D **96**(10), 103002 (2017). https://doi.org/10.1103/PhysRevD.96.103002
24. F. Hasegawa, M. Kawasaki (2018)
25. A.G. Polnarev, M. Yu. Khlopov, Sov. Phys. Usp. **28**, 213 (1985).https://doi.org/10.1070/PU1985v028n03ABEH003858. [Usp.Fiz.Nauk145,369(1985)]
26. E. Cotner, A. Kusenko, V. Takhistov (2018)
27. G.M. Fuller, A. Kusenko, V. Takhistov, Phys. Rev. Lett. **119**(6), 061101 (2017). https://doi.org/10.1103/PhysRevLett.119.061101
28. C. Kouvaris, P. Tinyakov, Phys. Rev. D **90**(4), 043512 (2014). https://doi.org/10.1103/PhysRevD.90.043512
29. A.P. Ji, A. Frebel, A. Chiti, J.D. Simon, Nature **531**, 610 (2016). https://doi.org/10.1038/nature17425

30. V. Takhistov, Phys. Lett. B **782**, 77 (2018). https://doi.org/10.1016/j.physletb.2018.05.026
31. R. Takahashi, T. Nakamura, Astrophys. J. **595**, 1039 (2003). https://doi.org/10.1086/377430
32. N. Matsunaga, K. Yamamoto, JCAP **0601**, 023 (2006). https://doi.org/10.1088/1475-7516/2006/01/023

Versatile Physics with Liquid Xenon Dark Matter Detectors

Rafael F. Lang

Abstract The much-discussed neutrino floor from atmospheric neutrinos will limit the sensitivity to directly search for WIMP dark matter, but is currently still well beyond our capabilities, namely by three orders of magnitude in rate and two generations in detectors. Liquid xenon-based detectors designed to truly probe WIMPs across this parameter range are sensitive to a wide range of physics channels, ranging from dark matter to neutrino physics and touching particle physics, nuclear physics and astrophysics. This contribution puts the current state of the art into perspective and sketches the science that can be done with current and upcoming liquid xenon detectors.

Keywords WIMPs · Dark matter · Xenon · XENON1T · LZ · LBECA · Direct detection · Solar neutrinos · Supernova

1 Context

As evidenced by other contributions to this workshop, the allowed dark matter mass range spans some 80 orders of magnitude, and a multitude of couplings and interaction channels are possible. It is, thus, obvious that a prohibitively large number of experiments is needed to probe the allowed dark matter parameter range in its entirety. Given the current absence of detections, there is thus an urgent need for theoretical motivations in order to prioritize the available parameter space [1]. It is particularly in this context that the WIMP paradigm in all its facets still provides an extremely well-motivated case to search for particles with masses above a GeV or so [2, 3]. Example of interactions come from couplings through the Higgs field, through suppressed weak charges, or even a simple coupling through the Z-boson at loop level, to just name some of the most straightforward hypotheses.

R. F. Lang (✉)
Department of Physics and Astronomy, Purdue University, 525 Northwestern Av, West Lafayette, IN 47906, USA
e-mail: rafael@purdue.edu

© Springer Nature Switzerland AG 2019
R. Essig et al. (eds.), *Illuminating Dark Matter*, Astrophysics and Space Science Proceedings 56, https://doi.org/10.1007/978-3-030-31593-1_12

Increasingly, sensitive detectors are therefore being built to continue probing this WIMP parameter range. In extending their reach, these detectors also achieve better sensitivity to signals from other dark matter models such as axions and axion-like particles or dark photons. Furthermore, neutrino-induced signals of various origins become measurable. Even the search for neutrinoless double-beta decay becomes feasible using the same detectors. Thus, these experiments turn from single-purpose dark matter searches into observatories for particle, astroparticle and nuclear physics.

By far, the most successful detector technology to search for direct interactions of WIMPs particles is liquid xenon-based time projection chambers (TPCs), as pioneered by the XENON10 collaboration [4] and subsequently improved upon by the XENON100 [5], LUX [6], Panda-X [7] and XENON1T [8] collaborations. Future experiments such as LZ [9] and XENONnT [10] continue this programme, with a generation-3 detector, sometimes called DARWIN [11], on the horizon. Such a generation-3 detector will not only probe the entire accessible WIMP parameter space down to the signal from atmospheric neutrinos, but will also be a multichannel experiment with a versatile and exciting physics programme.

2 Liquid Xenon TPCs: Redundancy Is the Key to Success

For a direct dark matter search, the expected energies are very low, of order keV or even less, with a spectrum that is a falling exponential with energy. Thus, the sensitivity can be increased by going to the lowest energies allowed by the detector. As there is less energy deposited in the detector, the signals become smaller, and one simply runs out of information. This makes it hard to distinguish signal from background, where background is not only from environmental radioactivity but more importantly from instrumental artefacts and various processes happening at the quantum limit of the detector. In fact, over the past decade or more, none of the leading WIMP searches were truly limited in sensitivity by the expected background from environmental radioactivity, but rather by detector-specific artefacts such as dark counts, imperfect signal collection near surfaces or specific event topologies in the detector. Whereas, radioactive backgrounds are simulated ahead of time and care is taken in the construction of the experiment that they can be dealt with at the level required for the projected sensitivity, instrumental artefacts are much harder or impossible to predict and thus readily are the limiting factors. The key to extending the reach of a detector is therefore its ability to distinguish true dark matter-induced signals from other, often unexpected detector artefacts.

This challenge is ideally met by liquid xenon TPCs through a single, monolithic design, paired with redundant readout even at threshold. For an educational description of these experiments, the reader is referred to the Wikipedia articles on LUX [13] or XENON [12]. The monolithic design not only reduces the surface-to-volume ratio and thus the relevance of instrumental artefacts. Crucially, this large volume allows cross-checks that would otherwise be hard or even impossible. For example, neutron-induced nuclear recoils have long thought to be an indistinguishable signal in any

WIMP detector. This is not true for the large TPCs now being used: neutrons will tend to scatter multiple times. Thus, by measuring nuclear recoils with high multiplicity, one has an in situ measurement of the neutron flux and thus, via simulation, a measurement and limit on the allowed single scatter background from neutrons.

The second key feature of this technology lies in the redundancy of the extracted information from each event, even at threshold. For example, the vertical (z) coordinate of an event is measured by the drift time of the event, but in addition, it is also encoded in the width of the ionization ($S2$) peak, and even in the hit pattern of the scintillation light ($S1$). Requiring consistency between all these redundant parameters is a powerful tool to reject detector-specific artefacts, such as events happening in the gas phase, or low-energy events originating from accidental pile-up of individual photomultiplier dark counts and photoionization events, just to name some examples. Another simple example includes the fact that not only the sizes of both scintillation ($S1$) and ionization ($S2$) signals need to be consistent with a given signal (e.g. of low energy for a generic WIMP), but also their ratio has to satisfy that expected of nuclear (or electronic, depending on the model) recoils.

3 A Generation-3 Experiment Is Required or: The Atmospheric Neutrino Floor Is Far, Far Away

Recently, the process in which an incoming neutrino scatters off a target and induces a nuclear recoil was discovered [14]. Direct WIMP search experiments are scattering experiments and as such only sensitive to the transferred momentum. This leaves a degeneracy between the signal induced by heavy but slow WIMPs, on the one hand, and light but relativistic neutrinos via coherent scattering off the nucleus on the other. Thus, neutrino-induced signals from astrophysical sources can be plotted in the same parameter space as the usual WIMP limits [15].

Two such signals need to be discussed separately: One is that of solar boron-8 neutrinos which look similar to ~8 GeV WIMPs. This is an exciting signal that XENON1T or at the latest LZ and XENONnT will be able to measure. This signal will be the first measurement of neutrino physics using a dark matter detector with consequences for solar astrophysics and the solar metallicity problem. Concerning the search for WIMPs, this signal will be a welcome in situ calibration line at the lowest energies.

The other signal comes from atmospheric neutrinos. This signal is still about three orders of magnitude beyond the sensitivity of current experiments. Even the current-funded experiments LZ and XENONnT will not be able to probe the available WIMP parameter space down to this signal. Thus, there is a very strong motivation to pursue a larger and more sensitive liquid xenon detector. Such a generation-3 experiment will be able to probe the entire accessible WIMP parameter range down to the signal from atmospheric neutrinos while simultaneously providing opportunities for many other measurements, from other dark matter signals to neutrino physics including the search for neutrinoless double-beta decay.

4   From WIMPs to Dozens of Science Channels

Thanks to their low-energy threshold, low background, large exposures and interesting target material, liquid xenon TPCs more and more turned into versatile science machines. Low-energy nuclear recoils can be interpreted in terms of spin-independent [6–8] or spin-dependent [16] interactions; more generally using effective field theory [17] or more specifically assuming light mediators as in the case of self-interacting dark matter models [18]. Those searches can be pushed to reach lower masses [19, 20] or much higher masses reaching even up to the Planck scale [21].

The extreme self-shielding in liquid xenon reduces the low-energy background by four–five orders of magnitude. This makes even the electronic recoil background very interesting to search for signals from dark matter. A particular interesting newly proposed [22] channels is through the Migdal effect, where inelastic scattering from low-mass WIMPs results in an electronic recoil above threshold [23]. Searches have been published for axion-like particles [24], SuperWIMPs and dark photons [25] as well as solar axions [26], leptophilic WIMPs [27], bosonic superWIMPs [28], mirror dark matter [29] and WIMPs scattering inelastically off the xenon [30].

Further very interesting signals come from the neutrino sector. For one, there are the above-mentioned future astroparticle measurements of electronic recoils from pp solar neutrinos, nuclear recoils from solar boron-8 neutrinos, as well as atmospheric neutrinos [15]. Neutrinos also provide the means to probe other physics beyond the standard model, for example through signals from sterile neutrinos or neutrino magnetic moments [31]. See, in particular, the contribution by Roni Harnik on chapter 'Interplay of Dark Matter Direct Detection and Neutrino Experiments' for more on these possibilities. Should a supernova goe off anywhere in our Milky Way, already running xenon experiments are sensitive to the nuclear recoils from such supernova neutrinos [32]. Double-electron capture has been searched for using the XMASS detector [33] with a sensitivity that can be improved upon with XENON1T. Finally, future liquid xenon dark matter detectors will also be competitive to search for neutrinoless double-beta decay [11].

5   The LBECA Approach

Thermal relic particles in the MeV–GeV mass range are another interesting target. Directly probing this mass range has been pioneered using LXe detectors [34] but hit previously ignored background sources at the level of individual or several electrons. While some sources of these few-electron backgrounds have been identified, others remain mere hypotheses, and methods to eliminate them are lacking. The LBECA collaboration has been formed to tackle this issue by identifying the various backgrounds and reduce them through dedicated changes to the detector hardware. The goal is to realize a dedicated, small detector that can overcome the major background sources and will feature an improved sensitivity for such low-energy signals.

6 Outlook: A Request to Theory

There are two simple cases for the future of this field. One, a detection of dark matter particles would be established by the current suite of experiments. In this case, the path forward is clear, namely to measure the properties of the underlying particle, its velocity distribution, couplings, etc. The other path is the more interesting one to be thinking about now: What if the currently experimental suite will not identify a dark matter particle, even in the next decade? Which experiments should one design and build then? Which will be the most well-motivated candidates that should be probed, and how would one go about it?

The dark matter motivation is not going anywhere, and if anything, the identification of particle candidates is only becoming more pressing. Thus, irrespective of the technical obstacles that may preclude realization of a given proposed search in this decade, the motivation to build an appropriate experiment will become better stronger and stronger. New technologies should be expected to push the limits of what is currently possible, in turn requiring a significant period of R&D. This development can be anticipated today: In proposing new well-motivated particle hypotheses and techniques to test them, one can lay today the foundation of a desirable experimental programme for years to come.

References

1. M. Battaglieri et al., (2017). arXiv:1707.04591
2. J.L. Feng, Annu. Rev. Astron. Astrophys. **48**, 495 (2010). https://doi.org/10.1146/annurev-astro-082708-101659
3. R.K. Leane, T.R. Slatyer, J.F. Beacom, K.C.Y. Ng, Phys. Rev. D **98**(2), 023016 (2018). https://doi.org/10.1103/PhysRevD.98.023016
4. J. Angle et al., Phys. Rev. Lett. **100**, 021303 (2008). https://doi.org/10.1103/PhysRevLett.100.021303
5. E. Aprile et al., Phys. Rev. Lett. **109**, 181301 (2012). https://doi.org/10.1103/PhysRevLett.109.181301
6. D.S. Akerib et al., Phys. Rev. Lett. **112**, 091303 (2014). https://doi.org/10.1103/PhysRevLett.112.091303
7. X. Cui et al., Phys. Rev. Lett. **119**(18), 181302 (2017). https://doi.org/10.1103/PhysRevLett.119.181302
8. E. Aprile et al., Phys. Rev. Lett. **121**, 111302 (2018). https://doi.org/10.1103/PhysRevLett.121.111302
9. B.J. Mount et al., (2017). arXiv:1703.09144
10. E. Aprile et al., JCAP **1604**(04), 027 (2016). https://doi.org/10.1088/1475-7516/2016/04/027
11. J. Aalbers et al., JCAP **1611**, 017 (2016). https://doi.org/10.1088/1475-7516/2016/11/017
12. Wikipedia, XENON, https://en.wikipedia.org/wiki/XENON
13. Wikipedia, Large underground xenon experiment. https://en.wikipedia.org/wiki/Large_Underground_Xenon_experiment
14. D. Akimov et al., Science **357**(6356), 1123 (2017). https://doi.org/10.1126/science.aao0990
15. J. Billard, L. Strigari, E. Figueroa-Feliciano, Phys. Rev. D **89**(2), 023524 (2014). https://doi.org/10.1103/PhysRevD.89.023524

16. D.S. Akerib et al., Phys. Rev. Lett. **116**(16), 161302 (2016). https://doi.org/10.1103/PhysRevLett.116.161302
17. J. Xia et al., Phys. Lett. B **792**, 193 (2019). https://doi.org/10.1016/j.physletb.2019.02.043
18. X. Ren et al., Phys. Rev. Lett. **121**(2), 021304 (2018). https://doi.org/10.1103/PhysRevLett.121.021304
19. E. Aprile et al., Phys. Rev. D **94**(9), 092001 (2016). https://doi.org/10.1103/PhysRevD.94.092001, https://doi.org/10.1103/PhysRevD.95.059901. [Erratum: Phys. Rev. D **95**(5), 059901 2017)]
20. R. Essig, T. Volansky, T.T. Yu, Phys. Rev. D **96**(4), 043017 (2017). https://doi.org/10.1103/PhysRevD.96.043017
21. J. Bramante, B. Broerman, R.F. Lang, N. Raj, Phys. Rev. D **98**(8), 083516 (2018). https://doi.org/10.1103/PhysRevD.98.083516
22. M.J. Dolan, F. Kahlhoefer, C. McCabe, Phys. Rev. Lett. **121**, 101801 (2018). https://doi.org/10.1103/PhysRevLett.121.101801
23. M. Ibe, W. Nakano, Y. Shoji, K. Suzuki, JHEP **03**, 194 (2018). https://doi.org/10.1007/JHEP03(2018)194
24. D.S. Akerib et al., Phys. Rev. Lett. **118**(26), 261301 (2017). https://doi.org/10.1103/PhysRevLett.118.261301
25. H. An, M. Pospelov, J. Pradler, A. Ritz, Phys. Lett. B **747**, 331 (2015). https://doi.org/10.1016/j.physletb.2015.06.018
26. E. Aprile et al., Phys. Rev. D **90**(6), 062009 (2014). https://doi.org/10.1103/PhysRevD.90.062009, https://doi.org/10.1103/PhysRevD.95.029904. [Erratum: Phys. Rev. D **95**(2), 029904 (2017)]
27. E. Aprile et al., Science **349**(6250), 851 (2015). https://doi.org/10.1126/science.aab2069
28. E. Aprile et al., Phys. Rev. D **96**(12), 122002 (2017). https://doi.org/10.1103/PhysRevD.96.122002
29. J.D. Clarke, R. Foot, Phys. Lett. B **766**, 29 (2017). https://doi.org/10.1016/j.physletb.2016.12.047
30. E. Aprile et al., Phys. Rev. D **96**(2), 022008 (2017). https://doi.org/10.1103/PhysRevD.96.022008
31. R. Harnik, J. Kopp, P.A.N. Machado, JCAP **1207**, 026 (2012). https://doi.org/10.1088/1475-7516/2012/07/026
32. R.F. Lang, C. McCabe, S. Reichard, M. Selvi, I. Tamborra, Phys. Rev. D **94**(10), 103009 (2016). https://doi.org/10.1103/PhysRevD.94.103009
33. K. Abe et al., PTEP **2018**(5), 053D03 (2018). https://doi.org/10.1093/ptep/pty053
34. J. Angle et al., Phys. Rev. Lett. **107**, 051301 (2011). https://doi.org/10.1103/PhysRevLett.110.249901, https://doi.org/10.1103/PhysRevLett.107.051301. [Erratum: Phys. Rev. Lett. **110**, 249901 (2013)]

The Origin of Galaxy Scaling Laws in LCDM

Julio F. Navarro

Abstract It has long been recognized that tight relations link the mass, size, and characteristic velocity of galaxies. These scaling laws reflect the way in which baryons populate, cool, and settle at the center of their host dark matter halos; the angular momentum they retain in the assembly process; as well as the radial distribution and mass scalings of the dark matter halos. There has been steady progress in our understanding of these processes in recent years, mainly as sophisticated N-body and hydrodynamical simulation techniques have enabled the numerical realization of galaxy models of ever increasing complexity, realism, and appeal. These simulations have now clarified the origin of these galaxy scaling laws in a universe dominated by cold dark matter: these relations arise from the tight (but highly nonlinear) relations between (i) galaxy mass and halo mass, (ii) galaxy size and halo characteristic radius; and (iii) from the self-similar mass nature of cold dark matter halo mass profiles. The excellent agreement between simulated and observed galaxy scaling laws is a resounding success for the LCDM cosmogony on the highly nonlinear scales of individual galaxies.

1 Introduction

The current paradigm for structure formation envisions a Universe whose matter component is dominated by cold dark matter and whose recently accelerated expansion reflects the negative pressure of a mysterious form of "dark energy" that resembles Einstein's cosmological constant (Lambda, or "L", for short). The nature of the dark matter and the source of dark energy constitute our era's premier challenges to our understanding of the physical universe.

Unraveling the nature of dark matter, in particular, is widely seen as the most promising way to extend the well-established Standard Model of Particle Physics, and is one of the most cherished goals of contemporary Theoretical Physics. Detailed modeling of the cosmic microwave background (CMB) and of large-scale galaxy

J. F. Navarro (✉)

Physics and Astronomy, University of Victoria, Victoria, BC V8P 5C2, Canada

e-mail: jfn@uvic.ca

© Springer Nature Switzerland AG 2019

R. Essig et al. (eds.), *Illuminating Dark Matter*, Astrophysics and Space Science Proceedings 56, https://doi.org/10.1007/978-3-030-31593-1_13

clustering have led to a few widely accepted conclusions: dark matter is almost certainly non-baryonic (or behaved as such at the time of primordial nucleosynthesis); it dominates roughly 5:1 over normal, baryonic matter, and clusters on a wide range of scales, from galaxy superclusters to dwarf galaxies [1].

Averaged over large scales, dark matter is distributed throughout the Universe in a web-like structure that matches closely that expected to arise from gravitational amplification of Gaussian density fluctuations. The fluctuation amplitude dependence on scale is also well constrained, and is broadly consistent with that expected from nearly scale-free perturbations in a collisionless fluid with small or negligible thermal velocities; i.e., "cold dark matter" (L+CDM, or "LCDM", for short). These successes imply that, at least in the quasi-linear regime probed by scales larger than about a small galaxy group, any successful model of dark matter must be or behave like CDM.

On smaller scales, there are no observational probes of the linear power spectrum, and, therefore, the clues rely on the clustering of dark matter inferred from observations in the highly nonlinear regime of individual galaxies. Since the galaxy baryonic component often plays a substantial role on these scales, the evidence is indirect, the predictions rely heavily on numerical simulations, and the interpretation is often inconclusive. Indeed, a number of "challenges" to LCDM have been identified on dwarf galaxy scales that, although not seen as lethal to LCDM, have attracted keen attention from advocates of modifications to LCDM or even to our standard model of gravity [2].

The purpose of this contribution is to add to this discussion by assessing the health of the LCDM paradigm on the nonlinear scales of individual galaxies. I focus on observations on the scale of the "L_*" galaxies where the large majority of stars in the universe reside [3, 4]. In particular, I describe the origin of the Tully–Fisher relation (TFR) that links the rotation speed of a disc galaxy with its stellar/baryonic mass in LCDM, as this is a sensitive test of the predicted nonlinear clustering of cold dark matter.

2 The Tully–Fisher Relation

The Tully–Fisher relation [5] is a particularly useful galaxy scaling relation because it links with tight scatter, distance-dependent, and distance-independent quantities. Properly calibrated, the TFR can therefore be used as a secondary distance indicator to map out the cosmic flows in the Local Universe and to measure Hubble's constant.

The observed TFR has long challenged direct numerical simulations of disc galaxy formation in LCDM. Indeed, early work produced galaxies so massive and compact that their rotation curves were steeply declining and, at given galaxy mass, peaked at much higher velocities than observed [6, 7].

The rotation speed of a disc depends on its baryonic mass and size (which set the contribution of the luminous component to the circular speed), as well as on the dark mass contained within the disk radius. The latter depends on the radial mass

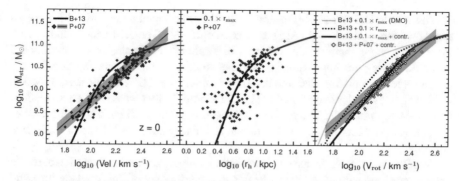

Fig. 1 Galaxy stellar mass, M_{str}, as a function of various parameters. *Left*: the solid black curve shows the abundance-matching prediction of [8], as a function of halo virial velocity, V_{200}. Symbols correspond to the data of [9], converted to stellar masses using a constant I-band mass-to-light ratio of 1.2 and shown as a function of disk rotation speed, V_{rot}. Color-shaded band indicates the mean slope and 1-σ scatter. *Middle*: symbols show half-light radii of galaxies in the P+07 sample. Thick solid line indicates a multiple of r_{max}, the characteristic radius where NFW halo circular velocities peak. Halo masses are as in the [8] model of the left panel. *Right*: Tully–Fisher relation. The color band is the same as in the left-hand panel. The dotted curve indicates the dark halo circular velocity at $r_h = 0.1\, r_{max}$, assuming NFW profiles and neglecting the contribution of the disk. The dashed line includes the gravitational contribution of the disk, keeping the halo unchanged. Finally, the thick solid line (and symbols) include the disk contribution *and* assume that halos contract adiabatically. This figure taken verbatim from [10]

Fig. 2 Tully–Fisher relation for EAGLE galaxies (gray band) compared with individual spirals taken from five recent TF compilations. The simulated relation is in excellent agreement with the observational data. The scatter is even smaller than in observed samples, even though the simulated relation includes *all* galaxies and not only disks. This figure taken verbatim from [10]

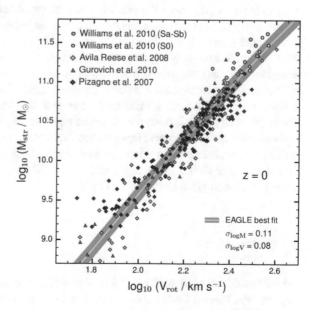

profile of dark matter halos, which is self-similar and well described in LCDM by the "NFW profile" [11]. The contribution of the dark matter to the circular velocity of a disc galaxy, then, depends only on the relation between galaxy mass and halo mass.

This relation is in turn fully constrained by the galaxy stellar mass function through the "abundance matching" (AM) approximation [8]. The AM ranks galaxies by mass and assigns them to halos ranked in similar fashion, preserving the ranked order. Satisfying this approximation appears to be a *sine qua non* condition for any cosmological simulation that attempts to reconcile the LCDM halo mass function with the galaxy stellar mass function [12]. This implies that there is no extra freedom in LCDM to "tune" the Tully–Fisher relation, making the TFR a useful probe of the clustering of dark matter distribution in the highly nonlinear scales of individual galaxies.

One feature of the AM approximation is that it predicts a complex relation between galaxy mass and halo mass. We show this in the left panel of Fig. 1, where the solid black line indicates the AM-derived halo virial[1] velocity (which is equivalent to halo virial mass; see X-axis) as a function of galaxy stellar mass (Y-axis). Disc rotation speeds for a sample of galaxies (the observed TFR) are shown by the symbols in the same panel. The observed relation clearly differs in shape and normalization from the AM relation between galaxy stellar mass and halo virial velocity, V_{200}.

However, disc rotation velocities are measured at the half-light radius, r_h, of the galaxy, and not at the virial radius. Half-light radii for the same galaxy sample are shown in the middle panel of Fig. 1. Measuring dark matter circular velocities at r_h leads to smaller values than V_{200} (gray curve on the left of the right-hand panel of Fig. 1) because, at r_h, the NFW halo circular velocity profile is still rising. This is even more at odds with the observed velocities. Adding the contribution of the baryonic disc, however, yields higher velocities, as indicated by the thick dotted line in the same panel. Finally, accounting for the response ("adiabatic contraction") of the halo to the assembly of the galaxy yields the thick solid line. This very crude model reproduces quite well the zero point and scatter of the TFR, as may be judged by the excellent agreement between the thick solid line and the symbols, which represent the model results when applied to the individual galaxies of the sample. The slope is slightly off from the observed relation, but this is a shortcoming of the approximate model adopted to represent the halo contraction. Indeed, a cosmological hydrodynamical simulation where galaxy disc masses roughly agree with AM and disc sizes agree with observation results in a TFR in excellent agreement with the observed relation [10], as shown by Fig. 2.

3   Outlook

We stress that the success of LCDM in accounting for the TFR is not simply a result of parameter tuning. Once the cosmological parameters are specified, if galaxies are assigned to halos so as to reproduce the galaxy stellar mass function and the galaxy

---

[1]The virial mass of a halo, M_{200}, is conventionally defined as that enclosed within a radius, r_{200}, where the mean density is 200 times $\rho_{crit} = 3H_0^2/8\pi G$, the critical density for closure. Virial quantities are measured at that radius and are listed with a "200" subscript.

mass–size relation is roughly in agreement with observation, then the resulting mass–velocity scaling for disc galaxies matches the observed TFR strikingly well. In other words, CDM halos add "just the right amount" of dark matter to the luminous regions of galaxies so as to reproduce the TFR. This is a nontrivial result that should rightfully be regarded as a true success of the LCDM cosmogony.

Key to this success is the self-similar "NFW" mass profile of LCDM. This profile implies that galaxies form in regions where the circular velocity of the halo is steadily rising, and where dark matter contribute a sizable, but not dominant, fraction of the mass enclosed within the half-light radius. The NFW profile shape is responsible for the rather small dispersion of the TFR: galaxies of different mass and size that populate halos of a given mass spread *along* the TFR, thus minimizing the scatter. We conclude that the TFR is a sensitive and telling test of the predicted clustering of CDM on the highly nonlinear scales corresponding to the half-light radii of disc galaxies. LCDM passes this test with flying colors.

Other galaxy scaling laws can also be used to test the predicted structure of LCDM halos. One example is the mass discrepancy–acceleration relation (MDAR), which links the spatial distribution of baryons with the speed/acceleration at which they orbit in galaxy discs [13]. This has also been examined in LCDM by a number of authors, who converge to conclude that the MDAR is just a reflection of the self-similar nature of cold dark matter halos and of the physical scales introduced by the galaxy formation process [14].

We have not examined here some of the small-scale challenges to LCDM high-lighted in other work, and expertly reviewed by [2]. These include the "missing satellites" and "rotation curve diversity" problems, the "too-big-to-fail" puzzle, the "missing dark matter galaxies", and the "cusp-core" controversy. We have addressed several of them in recent contributions, including [15–18], and have argued that all of them admit plausible resolutions in LCDM. The LCDM paradigm thus seems in excellent health, and news of its demise will, in the opinion of this author, prove exaggerated.

References

1. Planck Collaboration, N. Aghanim, Y. Akrami, M. Ashdown, J. Aumont, C. Baccigalupi, M. Ballardini, A.J. Banday, R.B. Barreiro, N. Bartolo, S. Basak, R. Battye, K. Benabed, J.P. Bernard, M. Bersanelli, P. Bielewicz, J.J. Bock, J.R. Bond, J. Borrill, F.R. Bouchet, F. Boulanger, M. Bucher, C. Burigana, R.C. Butler, E. Calabrese, J.F. Cardoso, J. Carron, A. Challinor, H.C. Chiang, J. Chluba, L.P.L. Colombo, C. Combet, D. Contreras, B.P. Crill, F. Cuttaia, P. de Bernardis, G. de Zotti, J. Delabrouille, J.M. Delouis, E. Di Valentino, J.M. Diego, O. Doré, M. Douspis, A. Ducout, X. Dupac, S. Dusini, G. Efstathiou, F. Elsner, T.A. Enßlin, H.K. Eriksen, Y. Fantaye, M. Farhang, J. Fergusson, R. Fernandez-Cobos, F. Finelli, F. Forastieri, M. Frailis, E. Franceschi, A. Frolov, S. Galeotta, S. Galli, K. Ganga, R.T. Génova-Santos, M. Gerbino, T. Ghosh, J. González-Nuevo, K.M. Górski, S. Gratton, A. Gruppuso, J.E. Gud-mundsson, J. Hamann, W. Handley, D. Herranz, E. Hivon, Z. Huang, A.H. Jaffe, W.C. Jones, A. Karakci, E. Keihänen, R. Keskitalo, K. Kiiveri, J. Kim, T.S. Kisner, L. Knox, N. Krach-malnicoff, M. Kunz, H. Kurki-Suonio, G. Lagache, J.M. Lamarre, A. Lasenby, M. Lattanzi,

C.R. Lawrence, M. Le Jeune, P. Lemos, J. Lesgourgues, F. Levrier, A. Lewis, M. Liguori, P.B. Lilje, M. Lilley, V. Lindholm, M. López-Caniego, P.M. Lubin, Y.Z. Ma, J.F. Macías-Pérez, G. Maggio, D. Maino, N. Mandolesi, A. Mangilli, A. Marcos-Caballero, M. Maris, P.G. Martin, M. Martinelli, E. Martínez-González, S. Matarrese, N. Mauri, J.D. McEwen, P.R. Meinhold, A. Melchiorri, A. Mennella, M. Migliaccio, M. Millea, S. Mitra, M.A. Miville-Deschênes, D. Molinari, L. Montier, G. Morgante, A. Moss, P. Natoli, H.U. Nørgaard-Nielsen, L. Pagano, D. Paoletti, B. Partridge, G. Patanchon, H.V. Peiris, F. Perrotta, V. Pettorino, F. Piacentini, L. Polastri, G. Polenta, J.L. Puget, J.P. Rachen, M. Reinecke, M. Remazeilles, A. Renzi, G. Rocha, C. Rosset, G. Roudier, J.A. Rubiño-Martín, B. Ruiz-Granados, L. Salvati, M. Sandri, M. Savelainen, D. Scott, E.P.S. Shellard, C. Sirignano, G. Sirri, L.D. Spencer, R. Sunyaev, A.S. Suur-Uski, J.A. Tauber, D. Tavagnacco, M. Tenti, L. Toffolatti, M. Tomasi, T. Trombetti, L. Valenziano, J. Valiviita, B. Van Tent, L. Vibert, P. Vielva, F. Villa, N. Vittorio, B.D. Wandelt, I.K. Wehus, M. White, S.D.M. White, A. Zacchei, A. Zonca (2018)

2. J.S. Bullock, M. Boylan-Kolchin, ARA&A **55**, 343 (2017). https://doi.org/10.1146/annurev-astro-091916-055313

3. C. Li, S.D.M. White, MNRAS **398**, 2177 (2009). https://doi.org/10.1111/j.1365-2966.2009.15268.x

4. I.K. Baldry, S.P. Driver, J. Loveday, E.N. Taylor, L.S. Kelvin, J. Liske, P. Norberg, A.S.G. Robotham, S. Brough, A.M. Hopkins, S.P. Bamford, J.A. Peacock, J. Bland-Hawthorn, C.J. Conselice, S.M. Croom, D.H. Jones, H.R. Parkinson, C.C. Popescu, M. Prescott, R.G. Sharp, R.J. Tuffs, MNRAS **421**, 621 (2012). https://doi.org/10.1111/j.1365-2966.2012.20340.x

5. R.B. Tully, J.R. Fisher, A&A **54**, 661 (1977)

6. J.F. Navarro, M. Steinmetz, ApJ **538**, 477 (2000). https://doi.org/10.1086/309175

7. C. Scannapieco, M. Wadepuhl, O.H. Parry, J.F. Navarro, A. Jenkins, V. Springel, R. Teyssier, E. Carlson, H.M.P. Couchman, R.A. Crain, C. Dalla Vecchia, C.S. Frenk, C. Kobayashi, P. Monaco, G. Murante, T. Okamoto, T. Quinn, J. Schaye, G.S. Stinson, T. Theuns, J. Wadsley, S.D.M. White, R. Woods, MNRAS **423**, 1726 (2012). https://doi.org/10.1111/j.1365-2966.2012.20993.x

8. P.S. Behroozi, R. Marchesini, R.H. Wechsler, A. Muzzin, C. Papovich, M. Stefanon, ApJ **777**, L10 (2013). https://doi.org/10.1088/2041-8205/777/1/L10

9. J. Pizagno, F. Prada, D.H. Weinberg, H.W. Rix, R.W. Pogge, E.K. Grebel, D. Harbeck, M. Blanton, J. Brinkmann, J.E. Gunn, AJ **134**, 945 (2007). https://doi.org/10.1086/519522

10. I. Ferrero, J.F. Navarro, M.G. Abadi, L.V. Sales, R.G. Bower, R.A. Crain, C.S. Frenk, M. Schaller, J. Schaye, T. Theuns, MNRAS **464**, 4736 (2017). https://doi.org/10.1093/mnras/stw2691

11. J.F. Navarro, C.S. Frenk, S.D.M. White, ApJ **490**, 493 (1997). https://doi.org/10.1086/304888

12. J. Schaye, R.A. Crain, R.G. Bower, M. Furlong, M. Schaller, T. Theuns, C. Dalla Vecchia, C.S. Frenk, I.G. McCarthy, J.C. Helly, A. Jenkins, Y.M. Rosas-Guevara, S.D.M. White, M. Baes et al., MNRAS **446**, 521 (2015). https://doi.org/10.1093/mnras/stu2058

13. S.S. McGaugh, F. Lelli, J.M. Schombert, Phys. Rev. Lett. **117**(20), 201101 (2016). https://doi.org/10.1103/PhysRevLett.117.201101

14. J.F. Navarro, A. Benítez-Llambay, A. Fattahi, C.S. Frenk, A.D. Ludlow, K.A. Oman, M. Schaller, T. Theuns, MNRAS **471**, 1841 (2017). https://doi.org/10.1093/mnras/stx1705

15. T. Sawala, C.S. Frenk, A. Fattahi, J.F. Navarro, R.G. Bower, R.A. Crain, C. Dalla Vecchia, M. Furlong, J.C. Helly, A. Jenkins, K.A. Oman, M. Schaller, J. Schaye, T. Theuns, J. Trayford, S.D.M. White, MNRAS **457**, 1931 (2016). https://doi.org/10.1093/mnras/stw145

16. A. Fattahi, J.F. Navarro, T. Sawala, C.S. Frenk, L.V. Sales, K. Oman, M. Schaller, J. Wang (2016)

17. K.A. Oman, J.F. Navarro, A. Fattahi, C.S. Frenk, T. Sawala, S.D.M. White, R. Bower, R.A. Crain, M. Furlong, M. Schaller, J. Schaye, T. Theuns, MNRAS **452**, 3650 (2015). https://doi.org/10.1093/mnras/stv1504

18. K.A. Oman, J.F. Navarro, L.V. Sales, A. Fattahi, C.S. Frenk, T. Sawala, M. Schaller, S.D.M. White, MNRAS **460**, 3610 (2016). https://doi.org/10.1093/mnras/stw1251

Searching for Light–Dark Matter with Positron Beams

Mauro Raggi

Abstract In the present contribution, we present preliminary ideas on the advantage of using positron beams on fixed target in searching for the A'. In particular, we will discuss A' production mechanisms, accessible only at positron machines, which might offer enhanced cross sections compared to the commonly used A'-strahlung. Recent studies show that the inclusion of such processes in the reinterpretation of old beam dump experiments results in significantly more stringent exclusion limits. Finally, we will discuss the peculiar case of searches for candidates with defined mass, using as a benchmark case the ^8Be 16.7 MeV X-Boson.

Keywords Dark photon · Resonant annihilation · Positron beam

1 Introduction

High-intensity positron beams have successfully been used as a source of A' at the B and ϕ factory experiments BaBar and KLOE. The luminosity achievable at e^+e^- colliders, limited the sensitivity to the coupling ϵ in the region down to ~ 1–2×10^{-3}. To explore lower values of the coupling, in the region between 10^{-3} and 10^{-5}, fixed-target experiments can profit from higher beam intensities, and high electrons density in solid materials. Fixed-target experiments searching for A' only used electron beams so far, limiting the accessible production mechanisms to A'-strahlung. In the past few years, several laboratories in the world have declared their interest in building high-intensity positron extracted beams [1]. It's therefore necessary to study in detail the pros and cons of using positrons in A' searches.

M. Raggi (✉)
Dipartimento di Fisica, Sapienza Università di Roma, Piazzale Aldo Moro 5, Rome, Italy
e-mail: mauro.raggi@roma1.infn.it

© Springer Nature Switzerland AG 2019
R. Essig et al. (eds.), *Illuminating Dark Matter*, Astrophysics and Space
Science Proceedings 56, https://doi.org/10.1007/978-3-030-31593-1_14

Fig. 1 A' production mechanisms using electron or positron beams [3]. **a** A'-strahlung in $e^{\pm}$-nucleon scattering; **b** A'-strahlung in e^+e^- annihilation; **c** resonant A' production in e^+e^- annihilation

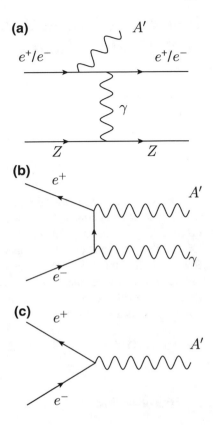

2 A' Production Using Electron or Positron Beams

The production of A' using electron beams has been, so far, the dominating mechanism in experimental searches at low energy. High intensity extracted electron beams have been successfully used in laboratories in US, and in Europe. The dark photon can be produced in collisions of electrons with a fixed target by the processes depicted in Fig. 1 diagram (a), analogous to ordinary photon bremsstrahlung, thanks to the kinetic mixing mechanism. The most commonly used expression for the differential cross section for A-strahlung in the Weizsäcker–Williams approximation reads [2]:

$$\frac{d\sigma}{dx\,d\cos\theta_{A'}} \approx \frac{8Z^2\alpha^3\epsilon^2 E_0^2 x}{U^2} \log\left[\left(1-x+\frac{x^2}{2}\right) - \frac{x(1-x)m_{A'}^2(E_0^2 x\theta_{A'}^2)}{U^2}\right], \quad (1)$$

dropping m_e and performing the angular integral:

$$\frac{d\sigma}{dx} \approx \frac{8Z^2\alpha^3\epsilon^2 x}{m_{A'}^2}\left(1+\frac{x^2}{3(1-x)}\right)\log, \quad (2)$$

The total rate scales by $(\alpha^3 \epsilon^2)/m_{A'}^2$. The $1/m_{A'}^2$ scaling of the cross section limits the A' mass reach of experiments based on A'-strahlung production to few hundreds of MeV. A critical discussion of limitations of the widely used Weizsäcker–Williams approximation in the region of $m_{A'} \approx 100$ MeV, shows that the production scales even faster if the exact calculation is performed [4].

Using positron beams, in addition to A-strahlung, two new contributions from annihilation processes, $(e^+ e^- \rightarrow \gamma A')$ Fig. 1b and $(e^+ e^- \rightarrow A')$ Fig. 1c, can significantly enhance the A' total production cross section.

The cross section for process (b), nonresonant annihilation, has the following expression:

$$\sigma_{nr} = \frac{8\pi\alpha^2}{s} \left[\left(\frac{s - m_{A'}^2}{2s} + \frac{m_{A'}^2}{s - m_{A'}^2} \right) \log \frac{s}{m_e^2} - \frac{s - m_{A'}^2}{2s} \right] \tag{3}$$

where s is the $e^+ e^-$ system invariant mass squared. This process allows to reach much higher values of $m_{A'}$, compared to A'-strahlung, especially at high-energy colliders. Babar at SLAC and KLOE at DAΦNE produced very strong constraints, based on nonresonant annihilation production, for both "visible" and "invisible" dark photon decays.

Just very recently the possibility of producing A' by resonant annihilation, Fig. 1c, has been put forward [5]. Given that $\epsilon \ll 1$ implies $\Gamma_{A'} = 1/3 m_{A'} \alpha \epsilon^2 \ll m_{A'}$, the cross section has been calculated, in the narrow-width approximation, to be [5]:

$$\sigma_r = \sigma_{peak} \frac{\Gamma_{A'}^2/4}{(\sqrt{s} - m_{A'})^2 + \Gamma_{A'}^2/4} = \frac{12\pi^2}{m_{A'}^2} \frac{\Gamma_{A'}^2/4}{(\sqrt{s} - m_{A'})^2 + \Gamma_{A'}^2/4} \tag{4}$$

The A' width is strongly suppressed by the small value of ϵ, and therefore the enhancement of the cross section is limited to a very narrow region of mass, preventing this mechanism from being used to explore wide parameter space at colliders. Nevertheless, its importance should not be overlooked for fixed-target experiment performed with variable energy positrons beams like PADME at the Laboratori Nazionali di Frascati.

3 Using Positrons in A' Searches

Several possibilities of exploiting these production techniques will be discussed in the following section, including the effect on the reinterpretation of existing results, the use of positrons in future beam dump experiments, and the opportunity that these production mechanisms will open for the upcoming PADME experiment at INFN Laboratori Nazionali di Frascati.

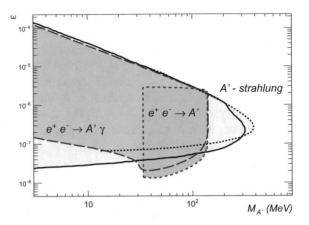

Fig. 2 A' exclusion limits from E137 considering e^+ nonresonant (long-dashed red line) and resonant (short-dashed blue line) production. Results from previous analysis which included only production via A'-strahlung are depicted as black-solid and black-dotted lines [6]

3.1 Effect of Positron Annihilation Production in Past Beam Dump Experiments

Electron beam dump experiments performed during the 80s produced an important set of constraints on the production of long-lived particles. Recent reinterpretations of these results in the A' parameter space constitute the more stringent constraints in the region of $\epsilon < 1 \times 10^{-4}$ values. It has been recently pointed out that electromagnetic showers are very rich in positrons, and, as a consequence, in recasting old beam dump experiment, the effect of the secondary positron induced A' production needs to be accounted for [3]. As a study case, the recast of the E137 experiment has been considered. The exclusion limit obtained, shown in Fig. 2, push down by a factor of two the exclusion region in the $m_{A'}$ range (35–120 MeV/c^2), implying that secondary positron annihilation needs to be included for a correct evaluation of exclusion limits obtained by electron beam dump experiments.

3.2 Targeting ^8Be with Resonant Production at PADME

A natural target for resonant production searches for defined mass states. This perspective is, in general, uncommon in the hunt for new physics, but an exception can arise from time to time. Recently, an anomalous pair production in the decays of excited ^8Be has been claimed, which might be explained by the existence of a new vector particle, so-called X-Boson, of ~17 MeV mass [7]. As pointed out in [5] the use of a ~282 MeV positron beam in dump mode will give to the PADME experiment the opportunity of producing the particle at resonance, providing a very peculiar enhancement of the signal, high energy e^+e^- pairs traversing a 10 cm W dump, around the production threshold. The potential exclusion limit is shown in Fig. 3a.

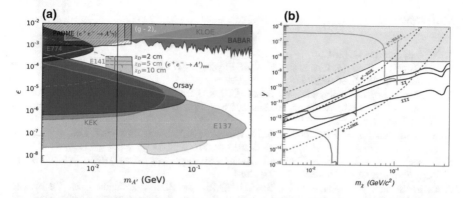

Fig. 3 **a** The three trapezoidal-shaped areas give the PADME reach in thick target resonant mode, respectively for a 10, 5 and 2 cm tungsten dump, assuming zero background [5]. **b** Continuous lines show exclusion limits at 90% C.L. for positrons beam dump experiments due to resonant and non-resonant positron annihilation (only). Dashed lines show exclusion limits obtained by considering A'-strahlung only [3]

Preliminary calculation also shows that a $\sim$282 MeV positron beam impinging on a thin diamond target will provide an enhancement in excess of 1000 in producing a 17 MeV vector particle, boosting the sensitivity of PADME for X-Boson searches.

3.3 Using Positrons in Future Beam Dump Experiments

In a similar way, the use of positrons in future beam dump and missing energy/momentum experiments can provide additional exclusion regions with respect to equivalent electron-based experiments.

In [3], exclusion limits for different future experiments, see Fig. 3b, are derived assuming the same (positron) charge and experimental efficiency quoted from original authors for the corresponding e^- beam setup. Significantly stronger constraints are obtained in specific region of the parameter space for all the different experimental techniques.

4 Conclusions

Positron beams constitute an important and competitive tool in searching for A'. Their use has been so far limited to experiments at colliders. New production mechanisms for A' production have been presented which might lead to significantly stronger exclusion limits compared to electron- based ones. Nowadays, positron beams with such characteristics are not available. Exploiting the described production mechanisms strongly motivates effort for the construction of intense positrons beam lines in the near future.

References

1. M. Battaglieri et al. (2017)
2. J.D. Bjorken, R. Essig, P. Schuster, N. Toro, Phys. Rev. D **80**, 075018 (2009). https://doi.org/10.1103/PhysRevD.80.075018
3. L. Marsicano, M. Battaglieri, M. Bondí, C.D.R. Carvajal, A. Celentano, M. De Napoli, R. De Vita, E. Nardi, M. Raggi, P. Valente, Phys. Rev. **D98**, 015031 (2018). https://doi.org/10.1103/PhysRevD.98.015031
4. S.N. Gninenko, D.V. Kirpichnikov, M.M. Kirsanov, N.V. Krasnikov, Phys. Lett. B **782**, 406 (2018). https://doi.org/10.1016/j.physletb.2018.05.010
5. E. Nardi, C.D.R. Carvajal, A. Ghoshal, D. Meloni, M. Raggi, Phys. Rev. D **97**(9), 095004 (2018). https://doi.org/10.1103/PhysRevD.97.095004
6. L. Marsicano, M. Battaglieri, M. Bondí, C.D.R. Carvajal, A. Celentano, M. De Napoli, R. De Vita, E. Nardi, M. Raggi, P. Valente, Phys. Rev. Lett. **121**, 041802 (2018). https://doi.org/10.1103/PhysRevLett.121.041802
7. A.J. Krasznahorkay et al., Phys. Rev. Lett. **116**(4), 042501 (2016). https://doi.org/10.1103/PhysRevLett.116.042501

Some Direct Detection Signatures
of Sub-MeV Dark Matter

Adam Ritz

Abstract Some of the motivations for considering light dark matter in the sub-GeV mass regime are reviewed, and I discuss work on specific direct detection signatures of MeV and sub-MeV mass dark matter candidates that can be studied using current experiments such as XENON1T, PANDA-X, SuperCDMS, and others.

1 Introduction

The WIMP paradigm emerged from the observation that a weakly interacting thermal relic–with an abundance determined by freeze-out in the early universe–naturally forms a cold dark matter candidate. Given that the WIMP annihilation rate scales as $\langle \sigma v \rangle \propto g^4 m_{DM}^2/m_{EW}^4$, viability of the scenario rests on the mass being above the Lee-Weinberg bound of a few GeV. WIMPs remain a compelling dark matter scenario, given this minimality. However, it is now well appreciated that the paradigm of a thermal relic species freezing out of thermal equilibrium with the Standard Model bath is more general [1–3], provided that there are additional forces beyond the weak interactions, so-called 'dark forces'. This leads to the idea of a multi-component 'dark sector', empirically motivated by both dark matter and neutrino mass, and there has been significant work exploring such hidden sectors experimentally over the past decade (see e.g. [4, 5]).

This generalization of the WIMP framework is still quite predictive, given a parametrization of the interactions between the SM and a hidden sector that assumes hidden sector states are SM gauge singlets. From the effective field theory expansion describing the interactions of light gauge singlet hidden sector fields with the SM, $\mathcal{L} \sim \sum_{n=k+l-4} \frac{c_n}{\Lambda^n} \mathcal{O}_{SM}^{(k)} \mathcal{O}_{hidden}^{(l)}$, it follows that the lower dimension interactions, namely those that are unsuppressed by the heavy scale Λ, are preferentially probed at lower energy. The set of relevant or marginal interactions, usually termed 'portals', is quite compact. Up to dimension four ($n \leq 0$), assuming SM electroweak symmetry breaking, the list of portals is as follows [6–9]:

A. Ritz (✉)
Department of Physics and Astronomy, University of Victoria, Victoria, BC, Canada
e-mail: aritz@uvic.ca

© Springer Nature Switzerland AG 2019
R. Essig et al. (eds.), *Illuminating Dark Matter*, Astrophysics and Space Science Proceedings 56, https://doi.org/10.1007/978-3-030-31593-1_15

Dark vectors	$-\frac{\epsilon}{2} B_{\mu\nu} F'^{\mu\nu}$
Dark scalars	$(AS + \lambda S^2) H^\dagger H$
Dark fermions	$y_N LHN$

where S, A'_μ and N are new SM-singlet degrees of freedom coupled to SM operators involving the Higgs double H, hypercharge field strength $B_{\mu\nu}$, and the LH singlet. On general grounds, the coupling constants for these interactions are unsuppressed by any heavy scale of new physics, and thus it would be natural for new weakly-coupled physics to first manifest itself via these portals. Indeed, we observe that the right-handed neutrino coupling is amongst this list, which provides the simplest renormalizable interpretation for neutrino mass and oscillations. It is natural to ask if the other portals are also realized, and in recent years all have been studied in the dark matter context (see e.g. [4, 5]).

While this framework is theoretically quite compact, it raises questions for direct detection, in which sensitivity normally weakens substantially for DM with a mass below a GeV, due to recoil energy thresholds and the reduced kinetic energy of the particle in the halo. This has motivated consideration of novel detection signatures, and also proposals for a range of new experiments. In this contribution, the focus is on reviewing a couple of novel signatures that can be analyzed with existing direct detection technology.

2 Direct Detection Signatures of Sub-MeV Mass DM

It is helpful to distinguish the parameter space of dark matter candidates in mass, and Fig. 1 illustrates the classical WIMP mass window and the full thermal relic mass range down to $m \sim m_e$, which is viable if there are additional dark forces to mediate annihilation. Below the electron mass threshold, it becomes more problematic to build thermal relic scenarios, but very weakly coupled dark matter candidates can arise through non-thermal 'freeze-in' production, e.g. via scattering off the SM bath. Generic candidates in both these categories can be detected using conventional direct detection experiments, as discussed below.

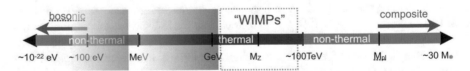

Fig. 1 The mass parameter range for dark matter, highlighting the thermal relic window in red, and the restricted range for conventional WIMPs. The shaded red and green regions refer to the scenarios considered in Sect. 2

2.1 Absorption

A general class of low mass (bosonic) dark matter candidates are too weakly coupled to the SM to thermalize in the early universe, and achieve a relic abundance through a variety of 'freeze-in' and other nonthermal production mechanisms. A prominent example is the QCD axion. However, another example which makes use of the renormalizable portals above is a massive dark photon, kinetically mixed with SM hypercharge. At energies well below the electroweak scale, the Lagrangian takes the form,

$$\mathcal{L} = -\frac{1}{4}F'^2_{\mu\nu} - \frac{\epsilon}{2}F'_{\mu\nu}F^{\mu\nu} + \frac{1}{2}m^2_{A'}A'^2_\mu. \tag{1}$$

Dark photon dark matter can be detected in conventional liquid xenon direct detection experiments, by inducing ionization if their mass exceeds the binding energy of electrons in the outermost shell, $m_{A'} > 12\,\text{eV}$. The cross section is directly analogous to the photoelectric effect,

$$\sigma_{A'}(E_{A'} = m_{A'})v_{A'} \simeq \epsilon^2\sigma_\gamma(\omega = m_{A'}). \tag{2}$$

This detection strategy was first discussed a decade ago in [10, 11], and the forthcoming sensitivity of XENON1T can push down to kinetic mixing values of $\epsilon \sim 10^{-16}$ as shown in Fig. 2 [12, 13].

Fig. 2 The sensitivity to very light dark photon dark matter, through absorption in liquid xenon [13], compared to a number of existing limits, e.g. from stellar cooling

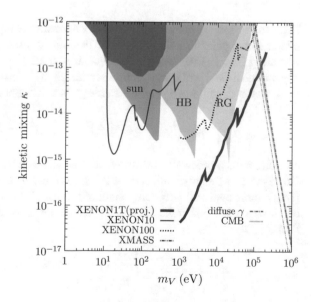

Fig. 3 The sensitivity of several direct detection experiments to sub-MeV dark matter due to solar reflection [15]

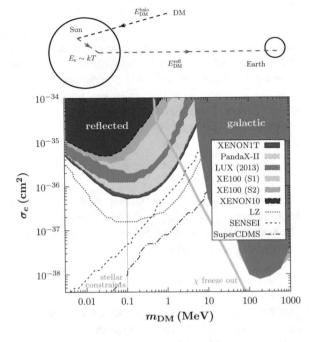

2.2 Solar Reflection and Electron Scattering

Direct detection experiments generally lose sensitivity rapidly once the mass of dark matter in the galactic halo falls below detector thresholds for energy deposited in scattering ($E_{\text{halo}} \sim 10^{-6} m_{\text{DM}}$). For nuclear scattering, this mass threshold is generally around a GeV, while for electron scattering the threshold is around 10 MeV [14]. Dark matter candidates χ, e.g. interacting via a dark photon,

$$\mathcal{L} = |(\partial_\mu - ie'A'_\mu)\chi|^2 - m_\chi^2|\chi|^2 - \frac{\epsilon}{2}F'_{\mu\nu}F^{\mu\nu}, \tag{3}$$

can have a mass below this threshold, and escape other direct or indirect constraints. However, even for dark matter below this mass threshold, as discussed in [15] (see also [16]) a more energetic sub-component of the halo DM flux induced by re-scattering off thermal electrons in the Sun can be detectable. This 'reflected' component of the DM flux Φ,

$$\Phi_{\text{refl}} \sim \frac{\Phi^{\text{halo}}}{4} \times \begin{cases} \frac{4S_g}{3}\left(\frac{R_{\text{core}}}{1\,\text{A.U.}}\right)^2 \sigma_e n_e^{\text{core}} R_{\text{core}}, & \sigma_e \ll 1\,\text{pb}, \\ S_g\left(\frac{R_{\text{scatt}}}{1\,\text{A.U.}}\right)^2, & \sigma_e \gg 1\,\text{pb}, \end{cases} \tag{4}$$

where σ_e is the scattering cross section on electrons and $S_g \sim 10$ is a gravitational focussing factor, satisfies $E_{\text{DM}}^{\text{refl}} < E_{\text{DM}}^{\text{refl,max}} = \frac{4E_e m_{\text{DM}} m_e}{(m_e + m_{\text{DM}})^2}$. This energy is generally above the ionization threshold for liquid xenon, and adds new sensitivity for any experiment able to detector electron ionization. The sensitivity of several experiments is shown in Fig. 3, and can be improved in the near future.

Acknowledgements I'd like to thank my colleagues, Haipeng An, Maxim Pospelov and Josef Pradler for enjoyable collaborations on the work described here, Rouven Essig, Jonathan Feng, and Kathryn Zurek for their invitation to this stimulating workshop, and the Simons Foundation for their financial support of the meeting. This work was supported in part by NSERC, Canada.

References

1. C. Boehm, P. Fayet, Nucl. Phys. B **683**, 219 (2004). https://doi.org/10.1016/j.nuclphysb.2004.01.015
2. J.L. Feng, J. Kumar, Phys. Rev. Lett. **101**, 231301 (2008). https://doi.org/10.1103/PhysRevLett.101.231301
3. M. Pospelov, A. Ritz, M.B. Voloshin, Phys. Lett. B **662**, 53 (2008). https://doi.org/10.1016/j.physletb.2008.02.052
4. J. Alexander et al. (2016), http://inspirehep.net/record/1484628/files/arXiv:1608.08632.pdf
5. M. Battaglieri et al., US Cosmic Visions: New Ideas in Dark Matter 2017: Community Report, arXiv:1707.04591 [hep-ph]
6. B. Holdom, Phys. Lett. **166B**, 196 (1986). https://doi.org/10.1016/0370-2693(86)91377-8
7. B. Patt, F. Wilczek, Higgs-field portal into hidden sectors, arXiv:hep-ph/0605188
8. R. Foot, H. Lew, R.R. Volkas, Phys. Lett. B **272**, 67 (1991). https://doi.org/10.1016/0370-2693(91)91013-L
9. B. Batell, M. Pospelov, A. Ritz, Phys. Rev. D **80**, 095024 (2009). https://doi.org/10.1103/PhysRevD.80.095024
10. M. Pospelov, A. Ritz, M.B. Voloshin, Phys. Rev. D **78**, 115012 (2008). https://doi.org/10.1103/PhysRevD.78.115012
11. J. Redondo, M. Postma, JCAP **0902**, 005 (2009). https://doi.org/10.1088/1475-7516/2009/02/005
12. H. An, M. Pospelov, J. Pradler, A. Ritz, Phys. Lett. B **747**, 331 (2015). https://doi.org/10.1016/j.physletb.2015.06.018
13. H. An, K. Ni, M. Pospelov, J. Pradler, A. Ritz, PoS **EPS-HEP2015**, 397 (2015)
14. R. Essig, A. Manalaysay, J. Mardon, P. Sorensen, T. Volansky, Phys. Rev. Lett. **109**, 021301 (2012). https://doi.org/10.1103/PhysRevLett.109.021301
15. H. An, M. Pospelov, J. Pradler, A. Ritz, Phys. Rev. Lett. **120**(14), 141801 (2018). https://doi.org/10.1103/PhysRevLett.120.141801
16. T. Emken, C. Kouvaris, N.G. Nielsen, Phys. Rev. D **97**(6), 063007 (2018). https://doi.org/10.1103/PhysRevD.97.063007

21 cm Absorption as a Probe of Dark Photons

Joshua T. Ruderman

Abstract Dark radiation could have injected soft photons into the primordial plasma with energies far below the Cosmic Microwave Background (CMB) temperature. Measurements of the low energy tail of the CMB spectrum therefore open a new window into the properties of dark radiation. We present an example model where dark radiation, composed of dark photons, resonantly oscillate into ordinary photons during the cosmic dark ages, enhancing the low energy tail of the CMB. Our scenario can explain the stronger than expected 21 cm absorption observed by the EDGES experiment.

1 Introduction

The spectrum of the Cosmic Microwave Background (CMB) has been measured to exquisite precision. FIRAS confirmed that the spectrum is consistent with a black-body at sub-mille precision and performed a precise measurement of its temperature, $T_{CMB} = 2.7255 \pm 0.0006$ K [1]. The FIRAS measurement relies on photon frequencies $\omega = 68$–639 GHz. Normalizing to the CMB temperature, $x \equiv \omega / T_{CMB}$, the FIRAS measurement corresponds to $x = 1.2$–11.2.

There are several measurements of the CMB at lower energies. ARCADE 2 measures the spectrum at $x = 0.056$ (where an excess is observed above the CMB temperature) [2]. Earlier measurements were conducted at $x \sim 0.02$–0.04 with larger uncertainties [3, 4]. Backgrounds become sizable at lower energies, and there exist no measurements of the CMB below $x = 0.01$.

As we review below, cosmological 21 cm absorption is sensitive to the number of CMB photons with wavelength 21 cm at redshifts of $z \approx 17$, corresponding to $x \approx 1.4 \times 10^{-3}$. The EDGES experiment has recently observed deeper 21 cm absorption than predicted by the standard cosmology, which can be explained if the CMB is enhanced at $x \sim 10^{-3}$.

J. T. Ruderman (✉)
Center for Cosmology and Particle Physics, Department of Physics, New York University, New York, NY 10003, USA
e-mail: ruderman@nyu.edu

© Springer Nature Switzerland AG 2019
R. Essig et al. (eds.), *Illuminating Dark Matter*, Astrophysics and Space Science Proceedings 56, https://doi.org/10.1007/978-3-030-31593-1_16

121

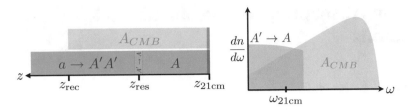

Fig. 1 An illustration of our mechanism. Dark matter decays to dark photons, $a \to A'A'$, which subsequently oscillate into ordinary photons, $A' \to A$. The dark photon spectrum is cut-off at the dark matter mass, $E_{A'} < m_a/2 \ll T_{\mathrm{CMB}}$, implying that the CMB is only enhanced at low energies

Below we describe an example model, illustrated in Fig. 1, where dark radiation composed of dark photons oscillate into ordinary photons. The resulting spectrum of ordinary photons is given by the expression,

$$\frac{dn_A}{d\omega} \to \frac{dn_A}{d\omega} \times P_{A \to A} + \frac{dn_{A'}}{d\omega} \times P_{A' \to A} \,, \tag{1}$$

where the $P_{A' \to A}$ describes the oscillation probability and $P_{A \to A} = 1 - P_{A' \to A}$ describes the survival probability of ordinary photons. As we describe below, our model produces an enhancement to the CMB at low energies, which can be probed by 21 cm, while leaving the bulk of the CMB spectrum unperturbed. For more details we refer the reader to Ref. [5]. A related scenario considers dark radiation composed of axion-like-particles that resonantly oscillate into soft photons in the presence of a primordial magnetic field [6].

2 21 cm Absorption as a CMB Thermometer

In the early Universe, 21 cm photons were absorbed or emitted as background photons passed through clouds of hydrogen gas (for reviews see Refs. [7, 8]). The brightness temperature of 21 cm absorption is given by [9],

$$\Delta T_{21}(z) = 32 \, \mathrm{mK} \times \left(1 - \frac{T_\gamma(z)}{T_s(z)}\right) \sqrt{\frac{1+z}{18}}, \tag{2}$$

where T_s is the hydrogen spin temperature, and T_γ counts the number of photons with wavelength 21 cm. In the standard cosmology, $T_\gamma = T_{\mathrm{CMB}}$, but more generally T_γ is proportional to the number of 21 cm photons.

At redshifts relevant to this discussion, the spin temperature is bounded below by the kinetic temperature of hydrogen gas, $T_s \geq T_k$. At $z \sim 20$, baryons are decoupled from the CMB and colder, $T_k < T_{\mathrm{CMB}}$, due to adiabatic cooling of the non-relativistic hydrogen gas. Lyman α photons from the first stars couple T_s to T_k, leading to

21 ̇cm absorption. The standard cosmology predicts $T_{21}(17) > -0.2$ K (where the lower bound corresponds to adiabatic cooling without heating), but the EDGES experiment observes $T_{21}(17) = -0.5^{+0.2}_{-0.5}$ K [10]. The depth of 21 cm absorption is enhanced if baryons are cooled by scattering with dark matter [11–13] or by earlier than expected decoupling from the CMB [14], although both of these possibilities are highly constrained by other measurements. Here, we consider the possibility that $T_\gamma > T_{\text{CMB}}$ (see also Ref. [15]), due to the injection of extra 21 cm photons from dark radiation.

3 Ordinary Photons from Dark Photons

We consider dark radiation composed of massive dark photons, A', that kinetically mix with ordinary photons,

$$\mathcal{L}_{AA'} = -\frac{1}{4}F_{\mu\nu}^2 - \frac{1}{4}(F'_{\mu\nu})^2 - \frac{\epsilon}{2}F_{\mu\nu}F'_{\mu\nu} + \frac{1}{2}m_{A'}^2(A'_\mu)^2 , \qquad (3)$$

where $\epsilon \ll 1$ is a dimensionless measure of the strength of kinetic mixing. Various bounds on dark photons are reviewed, as functions of ϵ and $m_{A'}$, by Ref. [16].

In the early universe, dark photons can oscillate into ordinary photons, $A' \to A$, and *vice versa*. While the probability of non-resonant oscillations is suppressed by ϵ^2, resonant oscillations occur when the the plasma mass of the ordinary photon crosses the mass of the dark photon [18]. The ordinary photon plasma mass (left of Fig. 2) is given by

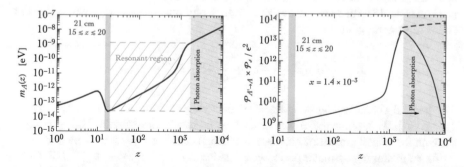

Fig. 2 The *left* panel shows the plasma mass of the ordinary photon, m_A, as a function of redshift, z. The plasma mass drops with redshift due to the expansion of the Universe and recombination. The *right* panel shows the oscillation probability, $P_{A'\to A}$ multiplied by the survival probability, P_s, normalized to the size of kinetic mixing, ϵ^2, as a function of the resonance redshift. These figures use the ionization fraction of Ref. [17]

$$m_A = \sqrt{\frac{4\pi\alpha n_e(z)}{m_e}} = 1.6 \times 10^{-14} \text{ eV} \times (1+z)^{3/2}\sqrt{x_e(z)}, \tag{4}$$

where n_e is the electron density and $x_e \equiv n_e/n_H$ describes the ionization fraction, with n_H the hydrogen density. The resonant oscillation probability (right of Fig. 2) is given by [18]

$$P_{A'\to A} = P_{A\to A'} = \frac{\pi\epsilon^2 m_{A'}^2}{\omega} \times \left|\frac{d\log m_A^2}{dt}\right|^{-1}_{t=t_{\text{res}}}. \tag{5}$$

Ordinary photons are abundantly produced at resonance. If photons with energy $x \sim 10^{-3}$ are produced at high redshifts, $z \gtrsim 1700$, they are rapidly absorbed by inverse Bremsstrahlung [19], and do not survive until 21 cm absorption occurs at $z \sim 15$–20. In order to impact the 21 cm absorption signal, resonance should occur between redshifts $z \sim 20$–1700, which requires a dark photon mass in the interval $m_{A'} = 10^{-14} - 10^{-9}$ eV. The resulting spectrum of ordinary photons is determined by the energy spectrum of dark photons at resonance, convolved with the resonance probability of Eq. 5.

4 Example Model with Dark Matter Decaying to Dark Photons

The final step is to specify the production mechanism of dark photons, which will determine their energy spectrum and the resulting spectrum of ordinary photons, after oscillation. We consider an example model where pseudoscalar dark matter, a, is metastable with a slow decay rate to dark photons, $a \to A'A'$. We assume the following Lagrangian,

$$\mathcal{L} \supset \frac{1}{2}(\partial_\mu a)^2 - \frac{m_a^2}{2}a^2 + \frac{a}{4f_a}F'_{\mu\nu}\tilde{F}'^{\mu\nu} + \mathcal{L}_{AA'}. \tag{6}$$

The general cosmological bound on DM decaying to dark radiation is $\tau_a \gtrsim 1.6 \times 10^{11}$ y [20]. We assume that a is at rest, as would result for example if it is produced through the misalignment mechanism, avoiding bounds on warm dark matter.

As an example parameter point, we consider $m_a = 10^{-3}$ eV and $f = 600$ GeV, which implies $\tau_a = 1.4 \times 10^{12}$ y. Figure 3 shows the spectrum of dark photons at this example point (blue curve) compared to the CMB spectrum (red curve) at $z = 500$. The dark photon spectrum is cut-off at high energies by the dark matter mass, $E_{A'} \leq m_a/2$. We also fix $\epsilon = 4 \times 10^{-7}$ and $m'_A = 5 \times 10^{-12}$ eV, which is consistent with existing bounds on dark photons [16]. This dark photon mass implies resonant oscillations, $A' \to A$, at $z \approx 500$. The purple curve of Fig. 3 shows the spectrum of ordinary photons produced at resonance. We see that the photons relevant for 21 cm absorption, $x \sim 10^{-3}$, are enhanced by an order one factor.

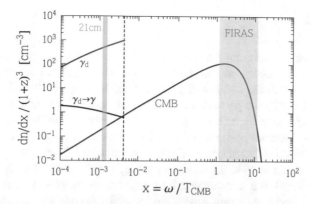

Fig. 3 The spectrum of dark photons (blue) and ordinary photons produced by $A' \rightarrow A$ resonance (purple), compared to the blackbody spectrum of CMB photons (red), evaluated at $z = 500$. We choose parameters $m_a = 10^{-3}$ eV, $\tau_a = 1.4 \times 10^{12}$ y, $\epsilon = 4 \times 10^{-7}$, and $m'_A = 5 \times 10^{-12}$ eV (corresponding to resonance at $z \approx 500$). The green region indicates photons with wavelength 21 cm at redshift $z = 15$–20, and the gray region indicates the photons measured precisely by FIRAS

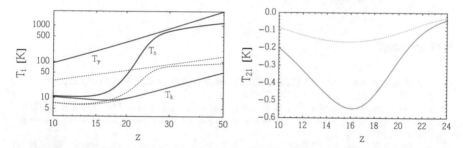

Fig. 4 The *left* panel shows the evolution of the photon, baryon spin, and baryon kinetic temperatures, T_γ, T_s, and T_k, as a function of redshift, z. The *right* panel shows the size of 21 cm absorption, T_{21}. In both panels, solid lines correspond to the example point in our model from Fig. 3 and dotted lines correspond to the standard cosmology. To produce this result we use the formalism of Ref. [9]. Note that these curves only include irreducible sources of baryon heating and do not include X-ray heating, which may dominate at low redshifts and modify the shape of the absorption feature

Finally, Fig. 4 shows how the 21 cm absorption is impacted by the extra ordinary photons injected at this example parameter point. The left panel shows the evolution of the photon temperature and the hydrogen spin and kinetic temperatures, and the right panel shows the depth of the 21 cm absorption. Solid lines correspond to our model and dotted lines to the standard cosmology. In order to produce these plots we use the formalism of Ref. [9], which only includes the irreducible sources of baryon heating. In particular, we do not include X-ray heating, which may dominate at low redshifts and modify the shape of the absorption feature.

For a broader excursion in parameter space, and the discussion of other relevant constraints, we refer the reader to Ref. [5].

5   Outlook

We have presented an example model where dark radiation, composed of dark photons, resonantly oscillates into ordinary photons. This results in an enhancement to the low energy tail of the CMB, without changing the shape of the bulk of the CMB spectrum. The model we presented is just one example of this general framework, and the type of dark radiation, and its production mechanism, can be varied while preserving the essential phenomenology. The main idea is that there is significant room for new physics that manifests through soft photons. The tentative observation of 21 cm absorption by EDGES probes a new regime of the CMB with photon energies only 10^{-3} the CMB temperature. Future 21 cm measurements will both test the EDGES result, and open a new window into the physics of dark radiation.

Acknowledgements Thanks goes to the organizers and to the Simons Foundation for making this engaging workshop possible. We would also like to thank Maxim Pospelov, Josef Pradler, and Alfredo Urbano for collaborating on the work described in this section. This work is supported by NSF CAREER grant PHY-1554858.

References

1. D.J. Fixsen, E.S. Cheng, J.M. Gales, J.C. Mather, R.A. Shafer, E.L. Wright, Astrophys. J. **473**, 576 (1996). https://doi.org/10.1086/178173
2. D.J. Fixsen et al., Astrophys. J. **734**, 5 (2011). https://doi.org/10.1088/0004-637X/734/1/5
3. M. Bersanelli, G.F. Smoot, M. Bensadoun, G. de Amici, M. Limon, S. Levin, Astrophys. Lett. Commun. **32**, 7 (1995)
4. S.T. Staggs, N.C. Jarosik, D.T. Wilkinson, E.J. Wollack, Astrophys. Lett. Commun. **32**, 3 (1995)
5. M. Pospelov, J. Pradler, J.T. Ruderman, A. Urbano, Phys. Rev. Lett. **121**(3), 031103 (2018). https://doi.org/10.1103/PhysRevLett.121.031103
6. T. Moroi, K. Nakayama, Y. Tang, Phys. Lett. B **783**, 301 (2018). https://doi.org/10.1016/j.physletb.2018.07.002
7. S. Furlanetto, S.P. Oh, F. Briggs, Phys. Rep. **433**, 181 (2006). https://doi.org/10.1016/j.physrep.2006.08.002
8. J.R. Pritchard, A. Loeb, Rep. Prog. Phys. **75**, 086901 (2012). https://doi.org/10.1088/0034-4885/75/8/086901
9. T. Venumadhav, L. Dai, A. Kaurov, M. Zaldarriaga (2018). https://doi.org/10.1103/PhysRevD.98.103513
10. J.D. Bowman, A.E.E. Rogers, R.A. Monsalve, T.J. Mozdzen, N. Mahesh, Nature **555**(7694), 67 (2018). https://doi.org/10.1038/nature25792
11. J.B. Muñoz, A. Loeb (2018). https://doi.org/10.1038/s41586-018-0151-x
12. A. Berlin, D. Hooper, G. Krnjaic, S.D. McDermott (2018). https://doi.org/10.1103/PhysRevLett.121.011102
13. R. Barkana, N.J. Outmezguine, D. Redigolo, T. Volansky (2018). https://doi.org/10.1103/PhysRevD.98.103005
14. A. Falkowski, K. Petraki (2018). arXiv:1803.10096
15. C. Feng, G. Holder, Astrophys. J. **858**(2), L17 (2018). https://doi.org/10.3847/2041-8213/aac0fe
16. R. Essig, et al., in *Proceedings, 2013 Community Summer Study on the Future of U.S. Particle Physics: Snowmass on the Mississippi (CSS2013)*, Minneapolis, MN, USA, July 29–August 6, 2013. http://inspirehep.net/record/1263039/files/arXiv:1311.0029.pdf

17. K.E. Kunze, M.A. Vazquez-Mozo, JCAP **1512**(12), 028 (2015). https://doi.org/10.1088/1475-7516/2015/12/028
18. A. Mirizzi, J. Redondo, G. Sigl, JCAP **0903**, 026 (2009). https://doi.org/10.1088/1475-7516/2009/03/026
19. J. Chluba, Mon. Not. R. Astron. Soc. **454**(4), 4182 (2015). https://doi.org/10.1093/mnras/stv2243
20. V. Poulin, P.D. Serpico, J. Lesgourgues, JCAP **1608**(08), 036 (2016). https://doi.org/10.1088/1475-7516/2016/08/036

Some Minimal Cosmologies for Dark Sectors

Jessie Shelton

Abstract One generic possibility for the origins of dark matter is its production from an internally thermalized hidden sector, with little to no direct involvement of the Standard Model. Any theory that invokes such a thermal dark radiation bath has to address the question of how this dark radiation bath was initially populated in the early universe. Here, we study how the simplest and most robust cosmic histories for minimal hidden sectors inform the signals of hidden sector dark matter, and present some new targets for direct detection and other terrestrial experiments.

1 Introduction

An enormous diversity of models is capable of explaining the observed dark matter (DM) relic abundance of our universe. Among this vast space of possibilities, however, thermal weakly interacting massive particles (WIMPs) have dominated the experimental hunt for dark matter for several reasons. The WIMP scenario is minimal, predictive, and can address other problems of the standard model (SM), in particular, the hierarchy problem. Unfortunately, the lack of evidence for WIMPs to date has severely curtailed the allowed parameter space. Beyond WIMPs, WIMP-like particles, which freeze-out directly to the SM through a new beyond-the-SM mediator, retain many of the advantages of WIMPs, in particular, the direct link between the observed relic abundance and predicted signals in terrestrial experiments. An equally minimal possibility, however, is hidden sector "WIMPs": here, DM freezes out directly to other dark states, with little or no direct involvement of the SM [1–4]. Hidden sectors, i.e., a set of SM singlet fields with possibly rich self-interactions, offer a variety of novel approaches to long-standing problems of the SM, from baryogenesis to the hierarchy problem, while providing natural explanations for the lack of BSM signals in terrestrial experiments to date; it is also interesting to note that hidden sectors are a frequent prediction of string compactifications that contain light SM-like degrees of freedom (e.g., [5]).

J. Shelton (✉)
Department of Physics, University of illinois at Urbana-Champaign, Urbana, IL, USA
e-mail: sheltonj@illinois.edu

© Springer Nature Switzerland AG 2019

R. Essig et al. (eds.), *Illuminating Dark Matter*, Astrophysics and Space Science Proceedings 56, https://doi.org/10.1007/978-3-030-31593-1_17

129

Thermal relics that arise from an internally thermalized hidden sector present an interesting but often challenging class of targets for DM searches. On one hand, these particles have the great advantage that they generally live at energy scales that are accessible with current or foreseeable technology, thanks to the remaining echoes of the "WIMP miracle". On the other hand, the couplings between these particles and the SM are typically very small and unrelated to the observed relic abundance, resulting in a lack of clear targets for terrestrial experiments. Fortunately, the need to populate a dark sector in the early universe generally (though not universally) places some requirements on its nongravitational interactions. I will discuss here two minimal cosmic origin scenarios for dark sectors, where the dark sector is produced in the early universe through its couplings to the SM, and explore their consequences for observability. Specifically, renormalizable "portal" operators are a well-motivated and natural choice for the leading interaction between the SM and a hidden sector. The importance of these portal interactions was emphasized by Jonathan Feng elsewhere in this volume; here we will further develop their possible consequences for dark cosmology.

2 WIMPs Next Door

Perhaps, the most minimal cosmological history for a dark sector is for it to have strong enough interactions with the SM to ensure that the two sectors reach thermal equilibrium in the early universe. When the dark sector equilibrates with the SM through renormalizable interactions, this cosmological history has the further advantage that it is "UV-safe": renormalizable operators have no intrinsic scale, and thus the scattering rates they mediate depend on temperature as $\Gamma \sim T$, i.e., scattering becomes more important relative to $H \propto T^2/M_{Pl}$ at low temperatures. Requiring $\Gamma > H$ prior to DM freeze-out thus places a lower bound on the leading coupling between the SM and the dark sector as a function of the freeze-out temperature T_{fo}; the UV-safety of renormalizable interactions guarantees that this lower bound is independent of the as-yet-unknown details of reheating. We call DM that freezes out of a dark radiation bath in thermal equilibrium with the SM a "WIMP next door" [6]. WIMPs next door have a clearly defined, bounded, and thus predictive parameter space of interest, and in particular, have a potentially accessible lower limit on their signals in terrestrial experiments.

We consider two simple minimal reference models, each of which contains a fermionic DM particle that annihilates to pairs of dark mediators that are in turn coupled to the SM. First is a vector model, where the dark sector and the SM are kept in thermal equilibrium by the renormalizable coupling to the $B - L$ current, $\epsilon \sum_f Q_f \bar{f} \gamma^\mu f Z_{D\mu}$, where f is a SM fermion with $B - L$ charge Q_f, Z_D^μ is a dark vector boson, and the dimensionless parameter ϵ controls thermalization as well as all terrestrial signals. Second is a scalar model, where a scalar mediator S couples to the SM through a Higgs-portal interaction, $\frac{\epsilon}{2} S^2 |H|^2$. The scalar S obtains a vacuum

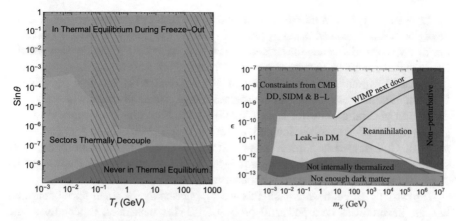

Fig. 1 Thermal coupling regions for scalar (left) and vector (right) models. Left: The orange region indicates where the dark sector is in thermal equilibrium with the SM at DM freeze-out temperature T_f, while in the blue region the two sectors were never in thermal equilibrium. In the green region, the SM and hidden sector were in equilibrium at some higher temperature, but fell out of equilibrium by T_f, so that the temperatures of the two sectors may drift apart. The hatched regions are near the chiral and electroweak phase transitions, where our calculations are less reliable. Right: thermalization floor (purple) as a function of DM mass, together with details of out-of-equilibrium evolution below. Figures from Refs. [6] and [7]

expectation value v_s and mixes with the SM Higgs, $\tan \theta \approx \frac{\epsilon v_h v_s}{m_h^2 - m_s^2}$, where s is the mass eigenstate in the symmetric vacuum. The mixing angle $\cos \theta$ controls all the relevant processes for either thermalization or discovery, except for those that proceed through an on-shell Higgs boson (e.g., exotic Higgs decays, which are the leading terrestrial discovery channel at masses $m_s \gtrsim 5\,\mathrm{GeV}$), which depends on a different function of ϵ and v_s. In both of these minimal models, the dark mediator furnishes the dark radiation bath; we thus require that it is relativistic at the time of DM freeze-out, which results in a cosmological lower bound on the portal coupling that is essentially independent of the mediator mass. In this regime the leading processes that the two sectors are $2 \leftrightarrow 2$ scattering processes, e.g., $gf \to Z_D f$. The resulting thermalization floors, shown in Fig. 1, are broadly representative of the thermalization floors for any vector- or Higgs-portal coupled dark radiation bath, with minimal modifications to account for the contribution of additional dark species to the heat capacity of the dark sector. However, we emphasize that it is the need to transfer enough energy to keep an entire radiation bath in thermal equilibrium with the SM that results in a terrestrially interesting value for the thermalization floor, and in models where all dark species are nonrelativistic at freeze-out, the thermalization constraint on the portal coupling is generally not restrictive compared to other observations.

In the regime of interest, the DM annihilation cross section depends only on couplings internal to the dark sector, not on the small portal couplings to the SM, and is fixed by the observed relic abundance. As the DM temperature cannot substantially differ from that of the SM, predictions from DM freeze-out are essentially identical to those from typical WIMP-like particles. In particular, DM masses for WIMPs next door cannot exceed $m_{DM} \lesssim$ few TeV, where the theory is weakly coupled; moreover, to avoid spoiling the successful predictions of BBN, both the DM and the dark mediator must be heavier than $\sim$ MeV. Indirect detection searches, especially in the CMB [8] and from gamma rays in dwarf galaxies [9], are now sensitive to DM annihilation cross sections in the thermal range, especially for lower mass DM, and are in general a powerful test of WIMPs next door. Our vector model has an s-wave annihilation cross section and, therefore, much of its parameter space below $m_\chi \lesssim 30$ GeV is ruled out. However, our scalar model has a p-wave annihilation cross section, as a simple and natural consequence of CP conservation, which renders its standard indirect detection signals unobservably small.

Terrestrially, direct detection signals are suppressed by the small portal coupling but can nevertheless be a powerful probe of WIMPs next door, especially in the regime where $m_{Z_D,s} \ll m_{DM}$. The thermalization floor determines a minimum possible direct detection cross section. Both vector and scalar portal interactions yield unsuppressed spin-independent nuclear scattering cross sections, and for both models *direct detection experiments are now probing well below the thermalization floor* in favorable regions of parameter space. In other regions of parameter space, signals are more challenging: for high DM masses and large values of $m_{Z_D,s}/m_{DM}$, the region of interest extends below the neutrino floor. The parameter space of interest for direct detection is summarized in Fig. 2. It is worth noting that the scalar model predicts a well-defined and potentially accessible target in the mass range MeV $\lesssim m_{DM} \lesssim 10$ GeV that can be of interest for new low-threshold direct detection experiments; in the vector model CMB constraints exclude this mass range.

Finally, the leading collider searches for WIMPs next door are for the dark mediators directly, with missing energy signals highly suppressed; exotic Higgs decays [18] are especially powerful probes of the scalar model.

3   Out-of-Equilibrium Interactions: Leak-In Dark Matter

An obvious question is: what happens for coupling strengths that are just below the thermalization floor? Couplings that do not quite equilibrate still allow the SM to transfer a macroscopic amount of energy into the HS, again allowing for the population of an internally thermalized hidden sector [19, 20], but now one that undergoes nonadiabatic evolution. A dark radiation bath being populated—i.e., freezing in—through an out-of-equilibrium renormalizable interaction reaches an attractor "leak-in" solution once the energy from the leak dominates over any initial abundance. In this leak-in phase, the energy density entering from the SM balances against the dilution from redshifting, yielding a temperature $\tilde{T}$ for the dark radiation bath that is related to the SM temperature T by [7]

$$\tilde{T}^4 = c\alpha_1 \epsilon^2 M_{Pl} T^3. \tag{1}$$

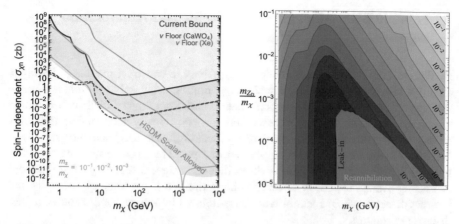

Fig. 2 Direct detection reach for Higgs portal (left) and $B - L$ (right) WIMPs next door. Left: The tan region indicates the allowed parameter space, while in green we show three lower bounds on the direct detection cross section for the mass ratios $m_s/m_\chi = 10^{-1}$, 10^{-2}, and 10^{-3}. The neutrino floors for xenon and CaWO$_4$ (used in CRESST) are shown with dashed purple and blue lines, respectively. The red line shows existing bounds from XENON1T [10], LUX [11, 12], PandaX-II [13], CDMSlite [14], and CRESST-II [15]. Right: Contours of maximum allowed portal coupling ϵ from Xenon1T [16], DarkSide-50 [17] and CRESST in the m_{Z_D}/m_χ vs. m_χ plane. The red region shows where the model is constrained to lie below the thermalization floor. Figures from Refs. [6] and [7]

Here c is a numerical constant that depends on the number of degrees of freedom in both sectors, α_1 is the fine structure constant for the SM interaction governing the freeze-in of the mediator abundance (e.g., for $gf \to Z_D f$, the relevant coupling is α_s), and ϵ is the small portal coupling. In particular, the energy density contained in this nonadiabatic radiation bath redshifts as $\rho_{HS} \propto a^{-3}$, like matter [7, 20].

As the leak-in phase is an attractor solution, it is a natural and generic possibility for DM freeze-out to occur while the hidden sector temperature follows $\tilde{T} \propto a^{-3/4}$. For freeze-out from this nonadiabatic radiation bath to yield the observed relic abundance, the necessary annihilation cross section is given approximately by [7]

$$\frac{\langle\sigma v\rangle_{\mathrm{LI}}}{\langle\sigma v\rangle_{\mathrm{WIMP}}} = b \left(\frac{\tilde{T}_{fo}}{m_{DM}}\right)^{1/3}, \tag{2}$$

where $b \propto \epsilon^{2/3}$ parameterizes the coldness of the hidden sector relative to the SM. The resulting indirect detection signals are thus suppressed compared to those expected for a "vanilla" thermal WIMP, making detection of these (simple, minimal) models extremely challenging over much of parameter space.

4 Possible Black Hole Probes of Feeble DM Annihilation Signals

As these examples indicate, it is depressingly easy to construct models of DM that are all but invisible to standard detection techniques, using only minimal and well-motivated ingredients in both particle content and cosmological history. It is worth noting that all of the models discussed here have cosmically mandated annihilation cross sections to visible final states. Broadly speaking, the indirect detection signals of these models are frequently less suppressed, compared to expectations for a "vanilla" thermal WIMP, than the collider or direct detection signals, and may offer the most promising avenue toward discovery. Nontraditional and exploratory strategies for indirect detection of DM annihilation constitute an important component of a DM discovery program.

One such exploratory strategy is provided by the DM density spikes that can form around supermassive black holes (SMBHs) [21]. Within the gravitational zone of influence of a SMBH, the dark matter distribution steepens into a power law spike, $\rho(r) \propto r^{-\gamma_{sp}}$, where r denotes the radius from the black hole, and the exponent γ_{sp} depends on the formation history of the black hole together with its halo, as well as the properties of its environment. (DM spikes can form around any point mass, but to yield annihilation signals in gamma rays at a level that is potentially interesting for DM models with suppressed annihilation rates, the steep, dense spikes that can form around SMBHs at the center of a DM halo are typically required.) The steepest spikes occur for BHs growing adiabatically at the center of cuspy host haloes, $\gamma_{sp} \approx 2.25 - 2.5$ [21], while BHs that form off-center generate only a very mild spike, $\gamma_{sp} = 0.5$ [22]. Other values between these two extremes can be realized with different formation histories for the BH and its host halo; see Ref. [23] for an overview.

In favorable scenarios, BH-induced density spikes can be enormously concentrated, making them highly sensitive probes of DM annihilation [21, 24]. This is thanks not only to the very large dark matter densities realized within such spikes but also to the increase in the DM velocity dispersion within the spike necessary to support the increased density, $v^2(r) \propto M_{BH}/r$. In other words, the SMBH acts as a mild DM accelerator, thereby potentially opening a late-universe door onto the physics of thermal freeze-out [25–28]. In particular, BH spikes can exclude the p-wave Higgs-portal WIMP next door model over a wide range of possible scenarios for the DM halo and spike distributions in the Galactic Center [28], and have excellent prospects for the faint s-wave annihilation signals of vector portal leak-in DM.

The search for DM annihilation in BH spikes is necessarily somewhat speculative: discovery is potentially achievable and would be spectacular, but null results have many possible explanations. For one thing, the center of the Milky Way is both busy and obscured, and the distribution of both stars and DM in the Galactic center has not been established to high precision; even if a spike is present, either it or the central halo may not be concentrated enough to yield interesting signals. Moreover, even if an interesting signal were to be detected, DM annihilations in BH spikes appears as a point like source to gamma ray telescopes. In the absence of the spatial information provided by extended halo signatures, sharp spectral features such as lines or boxes must be detected in order to be confident in ascribing a dark matter origin to any such point source. Third, numerous questions remain about the formation of black hole spikes in realistic galactic environments. Most spike solutions have been obtained in spherically symmetric contexts, starting from an idealized halo with a simple power law behavior, and their robustness in more realistic scenarios has yet to be conclusively shown. While the steepest adiabatic spikes are vulnerable to disruption, significant DM self-interactions [29] or significant gravitational scattering off stars [24, 30] can dynamically regenerate a spike and may lead to more durable features. The unique DM discovery handles potentially offered by BH spikes are excellent motivations both to pursue searches for spatially localized spectral features in cosmic rays and to further clarify the fate of spikes in realistic galactic environments.

Acknowledgements Many thanks to the organizers for putting together this stimulating workshop, and to the Simons Foundation for making it possible. Thanks also to my collaborators on the work highlighted here: J. Evans, B. Fields, C. Gaidau, S. Gori, and S. Shapiro. This work is supported in part by DOE Early Career award DE-SC0017840.

References

1. M. Pospelov, A. Ritz, M.B. Voloshin, Phys. Lett. B **662**, 53 (2008). https://doi.org/10.1016/j.physletb.2008.02.052
2. J.L. Feng, J. Kumar, Phys. Rev. Lett. **101**, 231301 (2008). https://doi.org/10.1103/PhysRevLett.101.231301
3. J.L. Feng, H. Tu, H.B. Yu, JCAP **0810**, 043 (2008). https://doi.org/10.1088/1475-7516/2008/10/043
4. N. Arkani-Hamed, D.P. Finkbeiner, T.R. Slatyer, N. Weiner, Phys. Rev. D **79**, 015014 (2009). https://doi.org/10.1103/PhysRevD.79.015014
5. A. Grassi, J. Halverson, J. Shaneson, W. Taylor, JHEP **01**, 086 (2015). https://doi.org/10.1007/JHEP01(2015)086
6. J.A. Evans, S. Gori, J. Shelton, JHEP **02**, 100 (2018). https://doi.org/10.1007/JHEP02(2018)100
7. J.A. Evans, C. Gaidau, J. Shelton, Leak-in dark matter. arXiv:1909.04671 [hep-ph]
8. P.A.R. Ade et al., Astron. Astrophys. **594**, A13 (2016). https://doi.org/10.1051/0004-6361/201525830
9. M. Ackermann et al., Phys. Rev. Lett. **115**(23), 231301 (2015). https://doi.org/10.1103/PhysRevLett.115.231301
10. E. Aprile et al., Phys. Rev. Lett. **119**(18), 181301 (2017). https://doi.org/10.1103/PhysRevLett.119.181301
11. D.S. Akerib et al., Phys. Rev. Lett. **118**(2), 021303 (2017). https://doi.org/10.1103/PhysRevLett.118.021303
12. D.S. Akerib et al., Phys. Rev. Lett. **116**(16), 161301 (2016). https://doi.org/10.1103/PhysRevLett.116.161301
13. X. Cui et al., Phys. Rev. Lett. **119**(18), 181302 (2017). https://doi.org/10.1103/PhysRevLett.119.181302
14. R. Agnese et al., Phys. Rev. Lett. **116**(7), 071301 (2016). https://doi.org/10.1103/PhysRevLett.116.071301

15. G. Angloher et al., Eur. Phys. J. C **76**(1), 25 (2016). https://doi.org/10.1140/epjc/s10052-016-3877-3
16. XENON collaboration, E. Aprile et al., Phys. Rev. Lett. **121** (2018) 111302. arXiv:1805.12562
17. DARKSIDE collaboration, P.Agnes et al., Phys. Rev. Lett. **121** (2018) 081307. arXiv:1802.06994
18. D. Curtin et al., Phys. Rev. D **90**(7), 075004 (2014). https://doi.org/10.1103/PhysRevD.90.075004
19. A.E. Faraggi, M. Pospelov, Astropart. Phys. **16**, 451 (2002). https://doi.org/10.1016/S0927-6505(01)00121-9
20. X. Chu, T. Hambye, M.H.G. Tytgat, JCAP **1205**, 034 (2012). https://doi.org/10.1088/1475-7516/2012/05/034
21. P. Gondolo, J. Silk, Phys. Rev. Lett. **83**, 1719 (1999). https://doi.org/10.1103/PhysRevLett.83.1719
22. P. Ullio, H. Zhao, M. Kamionkowski, Phys. Rev. D **64**, 043504 (2001). https://doi.org/10.1103/PhysRevD.64.043504
23. B.D. Fields, S.L. Shapiro, J. Shelton, Phys. Rev. Lett. **113**, 151302 (2014). https://doi.org/10.1103/PhysRevLett.113.151302
24. O.Y. Gnedin, J.R. Primack, Phys. Rev. Lett. **93**, 061302 (2004). https://doi.org/10.1103/PhysRevLett.93.061302
25. M.A. Amin, T. Wizansky, Phys. Rev. D **77**, 123510 (2008). https://doi.org/10.1103/PhysRevD.77.123510
26. M. Cannoni, M.E. Gomez, M.A. Perez-Garcia, J.D. Vergados, Phys. Rev. D **85**, 115015 (2012). https://doi.org/10.1103/PhysRevD.85.115015
27. C. Arina, T. Bringmann, J. Silk, M. Vollmann, Phys. Rev. D **90**(8), 083506 (2014). https://doi.org/10.1103/PhysRevD.90.083506
28. J. Shelton, S.L. Shapiro, B.D. Fields, Phys. Rev. Lett. **115**(23), 231302 (2015). https://doi.org/10.1103/PhysRevLett.115.231302
29. S.L. Shapiro, V. Paschalidis, Phys. Rev. D **89**(2), 023506 (2014). https://doi.org/10.1103/PhysRevD.89.023506
30. D. Merritt, Phys. Rev. Lett. **92**, 201304 (2004). https://doi.org/10.1103/PhysRevLett.92.201304

The SENSEI Experiment

Javier Tiffenberg

Abstract We present the status and prospects of the Sub-Electron-Noise Skipper Experimental Instrument (SENSEI) that uses a novel nondestructive readout technique to achieve stable readout for thick fully depleted silicon CCD in the far sub-electron regime ($\sim$0.05 e$^-$rms/pix). This is the first instrument to achieve discrete sub-electron counting that is stable over millions of pixels on a large-area detector. This low threshold allows for unprecedented sensitivity to the largely unexplored, but theoretically well-motivated, area of sub-GeV dark matter models. We will discuss the reach and prospects of the SENSEI experiment currently under construction, which will use 100 g of Skipper CCDs. We also present recent results from an engineering surface run and the lessons learned from a small scale prototype currently operating in the MINOS cavern at Fermilab.

1 Introduction

SENSEI is an exciting new experiment that fills an important hole in our search for DM. The objective of SENSEI is to search for a multitude of Hidden-Sector and Ultralight DM candidates with eV-to-GeV masses, see Fig. 1. The rationale for this objective is that such DM candidates are scientifically well-motivated but remarkably underexplored. Indeed, the wider DM community is recognizing the importance of searching for these non-WIMP DM candidates (for recent summaries see e.g. [1, 2]). SENSEI is complementary to searches for WIMPs by SuperCDMS, XENON1T, and LZ, and also complementary to searches for axion DM at much lower masses with ADMX and CASPEr.

SENSEI—the Sub-Electron-Noise Skipper CCD Experimental Instrument—uses silicon Charged Coupled Devices with a new, ultralow-noise readout technology ("Skipper CCDs"), designed in collaboration with the LBL MicroSystems Lab. In a recent technological breakthrough, SENSEI has demonstrated the ability to measure

J. Tiffenberg (✉)
Fermi National Accelerator Laboratory, Batavia, USA
e-mail: javiert@fnal.gov
URL: http://home.fnal.gov/~javiert/sensei/

© Springer Nature Switzerland AG 2019
R. Essig et al. (eds.), *Illuminating Dark Matter*, Astrophysics and Space
Science Proceedings 56, https://doi.org/10.1007/978-3-030-31593-1_18

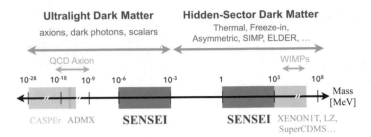

Fig. 1 SENSEI is dark matter (DM) direct detection experiment, which uses silicon CCDs with an ultralow-noise readout. It will search for Hidden-Sector DM with MeV-to-GeV masses that scatter off electrons and ultralight DM with eV-to-keV masses that is absorbed by electrons, probing several important DM candidates to unprecedented sensitivity

precisely the number of free electrons in each of the million pixels across the CCD [3]. The resulting detection threshold corresponds to 2 ionized electrons in a single pixel, compared to more than 10 ionized electrons in previous detectors. This enables a search of unprecedented sensitivity for MeV-to-GeV mass Hidden-Sector DM, which can scatter off an electron in the silicon and typically produces only a few ionized electrons [4, 5]; SENSEI can see these small signals, which are well below the threshold of previous detectors. Moreover, SENSEI can search for eV-to-keV mass Ultralight DM that is absorbed by an electron, probing to lower masses than ever before [6, 7].

2   The Major Advance Enabled by Skipper CCDs

The readout noise of previous silicon CCD detectors was about $2e^-$, requiring a threshold of $Q \geq 11e^-$ ($E_r \geq 40$ eV) [8]. Instead, the Skipper CCDs have a readout noise of $0.068e^-$, allowing precise and accurate measurement of the charge in each pixel, see Fig. 2 and [3].

In conventional scientific CCDs, low-frequency readout noise results in root mean squared (rms) variations in the measured charge per pixel at the level of $\sim 2e^-$ rms/pix [9, 10, and references therein]. In 1990 [11] proposed that low-frequency readout noise could be reduced by using a floating gate output stage [12] to perform repeated measurements of the charge in each pixel. This multiple readout technique was implemented in the form of a Skipper CCD [9, 11]; however, these early detectors suffered from spurious charge generation [13].

In 2017, [3] solved this problem by coupling the floating gate output stage of the Skipper CCD to a small capacitance sense node and isolating both from parasitic noise sources in order to perform multiple, independent, and nondestructive measurements of the charge in a single pixel . The result was a drastic reduction in low-frequency readout noise to the level of 0.068 rms/e$^-$ (as shown in Fig. 2). At this noise level, the

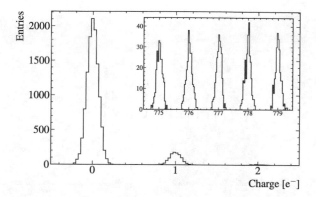

Fig. 2 Single-electron charge resolution using a Skipper CCD. The measured charge per pixel is shown for low (**main**) and high (**inset**) illumination levels. Integer electron peaks can be distinctly resolved in both regimes contemporaneously. The 0 e$^-$ peak has rms noise of 0.068 e$^-$rms/pix while the 777 e$^-$ peak has 0.086 e$^-$ rms/pix, demonstrating single-electron sensitivity over a large dynamical range. Plot is reproduced from [3]

probability p that the charge per pixel is misestimated by >0.5e$^-$ is $p \sim 10^{-13}$. This was the first accurate single-electron measurement in a large-format (4126×866 pix) silicon detector.

The low readout noise achieved by Skipper CCDs, coupled with a stable linear gain, allows charge measurement at the accuracy of individual electrons simultaneously in pixels with single electrons and thousands of electrons. This makes the Skipper CCD the most sensitive and robust electromagnetic calorimeter that can operate at temperatures above that of liquid nitrogen. Because nondestructive readout is achieved without any major modifications to the CCD fabrication process, this new technology can be immediately implemented in existing CCD manufacturing facilities at low cost.

3 Science Reach

The reach of the SENSEI detector for two well motivated electron-recoil DM candidates are summarized in Fig. 3. SENSEI will probe orders of magnitude of unexplored and important parameter space for two compelling classes of DM candidates beyond WIMPs: MeV-to-GeV masses for **Hidden-Sector DM** and eV-to-keV masses for **Ultralight DM**. While DM could hide anywhere in this parameter space, several sharp regions can be identified in which simple and motivated Hidden-Sector DM particles with ~MeV-to-GeV masses have the correct relic abundance from various production mechanisms in the early Universe. Direct-detection experiments are an essential laboratory tool to identify DM. The traditional technique is to search for ~10 keV-scale nuclear recoils characteristic of WIMPs with masses > 10 GeV scattering off nuclei in a detector [14]. Design and construction advances over the last few

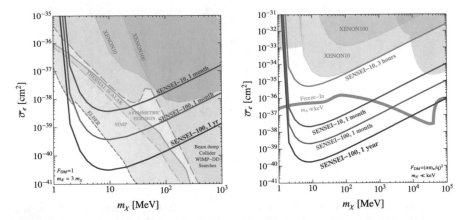

Fig. 3 Projections for SENSEI to probe Hidden-Sector Dark Matter that scatters off electrons. Blue regions show direct-detection constraints on electron recoils, from XENON10 and XENON100 [15, 18]. Gray regions show existing constraints from accelerator-based (LSND, E137, BaBar) or WIMP direct-detection searches [5]. **Left**: The DM scatters off electrons by exchanging a dark photon, A', with $m_{A'} = 3m_{\chi}$, leading to a momentum-independent interaction ($F_{DM}(q) = 1$). Four specific theory targets are shown, which assume different production mechanisms for the DM (thermal scalar (solid green line), asymmetric fermion (orange region), SIMP (cyan region), and ELDER (dark-cyan dashed line). **Right**: Here $m_{A'} \ll$ keV, leading to a momentum-dependent interaction with $\bar{\sigma}_e \propto |F_{DM}(q)|^2$ with $F_{DM}(q) = (\alpha m_e/q)^2$. The correct DM relic abundance is produced through the "freeze-in" mechanism along the solid brown line

decades have led to many orders of magnitude improvement in sensitivity. The traditional nuclear-recoil search is, however, unable to probe sub-GeV Hidden-Sector DM or sub-keV Ultralight DM, which give too little energy to the nucleus (below detector thresholds). Instead, a search for *low-energy electron recoils* can probe this important mass range [4–7, 15–22]. SENSEI's threshold of 2 ionized electrons is well below the threshold of previous detectors (> 10 ionized electrons) [3]. The implications of SENSEI's low threshold are profound, especially for the search of Hidden-Sector DM, which typically only produces a few ionized electrons, well below the energy threshold of previous detectors.

4   Surface Run Science Results

The unprecedented capabilities of the Skipper CCD detectors produced the first constraints on sub-GeV DM derived from a small set of SENSEI commissioning data taken at surface level with small matter overburden [23]. Using just 8 h of data taken with a small prototype detector it is possible to exclude novel parameter space for DM masses below ~4 MeV, above which the XENON10 constraint from [15, 18] dominates. Furthermore, operating on the surface allows a search for DM that strongly interacts with the visible sector. Such DM does not penetrate the Earth, and

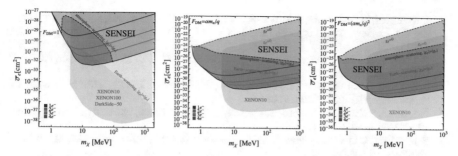

Fig. 4 The 95% C.L. constraints on the DM-electron scattering cross sections, $\overline{\sigma}_e$, as a function of DM mass, m_χ, from a commissioning run above ground at FNAL using the SENSEI prototype detector. We show different DM form factors, $F_{DM}(q) = 1$, $\alpha m_e/q$, and $(\alpha m_e/q)^2$. The purple, blue, green, and red lines correspond to the strongest constraints, from using events with exactly 1, 2, 3, or 4 electrons, respectively, with the black line indicating the strongest constraint for all DM masses. The blue shaded regions are the current constraints from DM-electron scattering from XENON10, XENON100, and DarkSide-50. For large cross sections, the DM is stopped in the Earth's crust (atmosphere) and does not reach the noble-liquid (SENSEI prototype) detectors: the dark-shaded regions (labeled $|g_p| = |g_e|$) show preliminary results from [24] and are the excluded parameter regions assuming the interaction between DM and ordinary matter is mediated by a heavy dark photon (left), an electric dipole moment (middle), or an ultralight dark photon (right). The light-shaded regions (labelled $g_p = 0$) are the approximate excluded parameter regions assuming a mediator that couples only to electrons. The terrestrial effects shown here are order-of-magnitude estimates only, and more detailed calculations will appear in [24]

detectors placed deep underground, such as the noble-liquid detectors mentioned above, have no sensitivity. Despite large cosmic-ray backgrounds, this region can be easily probed by a detector on the surface with a small amount of data. The SENSEI data thus also place novel constraints on DM particles with masses of several hundred MeV (Fig. 4).

5 Summary and Outlook

Over the next few years, the SENSEI Collaboration aims to construct a detector consisting of ~100 g of Skipper CCDs that are fabricated in a dedicated production run using high-resistivity silicon. We expect to collect an exposure that is almost 2 million times larger than the exposure of the surface run and with far fewer background events, allowing us to explore vast new regions of DM parameter space.

References

1. J. Alexander et al. (2016), http://inspirehep.net/record/1484628/files/arXiv:1608.08632.pdf
2. M. Battaglieri et al. (2017)

3. J. Tiffenberg, M. Sofo-Haro, A. Drlica-Wagner, R. Essig, Y. Guardincerri, S. Holland, T. Volansky, T.T. Yu, Phys. Rev. Lett. **119**(13), 131802 (2017). https://doi.org/10.1103/PhysRevLett.119.131802
4. R. Essig, J. Mardon, T. Volansky, Phys. Rev. D **85**, 076007 (2012). https://doi.org/10.1103/PhysRevD.85.076007
5. R. Essig, M. Fernandez-Serra, J. Mardon, A. Soto, T. Volansky, T.T. Yu, JHEP **05**, 046 (2016). https://doi.org/10.1007/JHEP05(2016)046
6. I.M. Bloch, R. Essig, K. Tobioka, T. Volansky, T.T. Yu, JHEP **06**, 087 (2017). https://doi.org/10.1007/JHEP06(2017)087
7. Y. Hochberg, T. Lin, K.M. Zurek, Phys. Rev. D **95**(2), 023013 (2017). https://doi.org/10.1103/PhysRevD.95.023013
8. J. Barreto et al., Phys. Lett. B **711**, 264 (2012). https://doi.org/10.1016/j.physletb.2012.04.006
9. J.R. Janesick, *Scientific Charge Coupled Devices* (SPIE Publications, Bellingham, 2001)
10. C. Bebek, J. Emes, D. Groom, S. Haque, S. Holland, P. Jelinsky, A. Karcher, W. Kolbe, J. Lee, N. Palaio, D. Schlegel, G. Wang, R. Groulx, R. Frost, J. Estrada, M. Bonati, J. Instrum. **12**(04), C04018 (2017), http://stacks.iop.org/1748-0221/12/i=04/a=C04018
11. C.E. Chandler, R.A. Bredthauer, J.R. Janesick, J.A. Westphal, J.E. Gunn, SPIE **1242**, 238 (1990)
12. D. Wen, IEEE J. Solid-State Circuits **9**(6), 410 (1974). https://doi.org/10.1109/JSSC.1974.1050535
13. G. Fernandez Moroni, J. Estrada, G. Cancelo, S.E. Holland, E.E. Paolini, H.T. Diehl, Exp. Astron. **34**, 43 (2012). https://doi.org/10.1007/s10686-012-9298-x
14. M.W. Goodman, E. Witten, Phys. Rev. D **31**, 3059 (1985). https://doi.org/10.1103/PhysRevD.31.3059
15. R. Essig, A. Manalaysay, J. Mardon, P. Sorensen, T. Volansky, Phys. Rev. Lett. **109**, 021301 (2012). https://doi.org/10.1103/PhysRevLett.109.021301
16. P.W. Graham, D.E. Kaplan, S. Rajendran, M.T. Walters, Phys. Dark Univ. **1**, 32 (2012). https://doi.org/10.1016/j.dark.2012.09.001
17. S.K. Lee, M. Lisanti, S. Mishra-Sharma, B.R. Safdi, Phys. Rev. D **92**(8), 083517 (2015). https://doi.org/10.1103/PhysRevD.92.083517
18. R. Essig, T. Volansky, T.T. Yu, Phys. Rev. D **96**(4), 043017 (2017). https://doi.org/10.1103/PhysRevD.96.043017
19. S. Derenzo, R. Essig, A. Massari, A. Soto, T.T. Yu (2016)
20. Y. Hochberg, Y. Zhao, K.M. Zurek, Phys. Rev. Lett. **116**(1), 011301 (2016). https://doi.org/10.1103/PhysRevLett.116.011301
21. Y. Hochberg, M. Pyle, Y. Zhao, K.M. Zurek, JHEP **08**, 057 (2016). https://doi.org/10.1007/JHEP08(2016)057
22. H. An, M. Pospelov, J. Pradler, A. Ritz, Phys. Lett. B **747**, 331 (2015). https://doi.org/10.1016/j.physletb.2015.06.018
23. M. Crisler, R. Essig, J. Estrada, G. Fernandez, J. Tiffenberg, M. Sofo Haro, T. Volansky, T.T. Yu (2018)
24. T. Emken, R. Essig, C. Kouvaris, M. Sholapurkar

Indirect Probes of Light Dark Matter

Tomer Volansky

Abstract So far, dark matter has only been discovered gravitationally, while its particle identity remains unknown. It is possible that dark matter is so weakly coupled to the visible sector that a direct nongravitational interaction lies well beyond our experimental reach. It is then interesting to ask to what extent indirect probes of dark matter can point to a specific particle physics description. In this note, we discuss two such examples: The first is via 21 cm cosmology and the second is via the study of AGN and black hole growth rate.

1 Introduction

Since the early twentieth century, physicists have been trying to identify the origin of dark matter. So far, however, dark matter has only been detected gravitationally, showing no hint of its particle identity. Indeed, going beyond the astrophysical and cosmological observations, there have been several directions to search for dark matter, directly with underground detectors, indirectly with the use of satellites and earth-based telescopes, and at colliders such as the LHC.

It is very possible that a nongravitational discovery of dark matter is just around the corner. Many motivated models point to dark matter properties that can be probed by ongoing and upcoming experiments. However, it is equally easy to envision dark matter models which would never be discovered nongravitationally. Perhaps the simplest and most known example is a scalar field that interacts only gravitationally, slowly redshifting as it oscillates around the minimum of its potential since the end of inflation.

In light of such theoretical possibilities, it is interesting to ask whether astrophysical and cosmological probes of dark matter, which do not directly observe dark matter interactions, can aid in pinpointing the particle identity of dark matter despite being indirect. Perhaps the most studied example of this kind is self-interacting dark

T. Volansky (✉)
Raymond and Beverly Sackler School of Physics and Astronomy, Tel-Aviv University, 69978 Tel Aviv-Yafo, Israel
e-mail: tomerv@post.tau.ac.il

© Springer Nature Switzerland AG 2019
R. Essig et al. (eds.), *Illuminating Dark Matter*, Astrophysics and Space Science Proceedings 56, https://doi.org/10.1007/978-3-030-31593-1_19

matter [1–3] where several small-scale discrepancies can be interpreted as hinting toward a specific form of dark matter.

In this note, we discuss two distinct examples of such indirect probes of dark matter. In the first [4], dark matter interactions affect either the CMB or gas temperatures during the dark ages ($z \sim 20$–150). Such interactions are motivated by the recent EDGES measurement [5] and we show that cooling the gas is an unlikely explanation. Although this example demonstrates an indirect probe of dark matter that can aid in identifying the particle properties, it requires additional nongravitation (strong) interactions which guarantee other avenues of discovery.

The second [6] example shows sensitivity to a truly secluded dark matter. In this case, we study the growth rate of Active Galactic Nuclei (AGN) focusing, as an example, on a set of measurements that hint on an anomalously fast growth rate. We show that these measurements can be explained by the presence of a dissipative dark matter component that can form a dark accretion disk.

We believe that these examples are only two of many that may allow to better understand the particle description of dark matter without a direct observation of its interactions. If dark matter is only feebly interacting with the visible sector, such studies may play a crucial role in our ongoing exploration of the universe.

2   Probing Dark Matter in 21 cm Cosmology

2.1   21cm Cosmology Basics and EDGES

Below $z \sim 200$ and down to $z \sim 30$, the baryonic gas (composed mostly of hydrogen) is decoupled from radiation and cools adiabatically. At this time, most of the hydrogen gas is in its ground state, whose degeneracy is only broken by the hyperfine splitting with an energy difference of $E_{21} = 5.9 \times 10^{-6}$ eV $\simeq 0.068$ K $\simeq 2\pi/21$ cm. The relative number density of triplet and singlet states of the hydrogen defines the so-called spin temperature,

$$\frac{n_1}{n_0} \equiv \frac{g_1}{g_0} e^{-E_{21}/T_s} \simeq 3 \left(1 - \frac{E_{21}}{T_s} \right) . \tag{1}$$

This effective temperature is sensitive to different spin-flipping processes after recombination, in particular: H-H and H-e collisions, resonant absorption of CMB radiation, and scattering of UV photons. Solving the Boltzmann equations one finds,

$$\Delta T_s \simeq \frac{y_{\text{col}} \, \Delta T_{\text{gas}} + y_{\text{Ly}\alpha} \, \Delta T_{\text{Ly}\alpha}}{1 + y_{\text{col}} + y_{\text{Ly}\alpha}} , \tag{2}$$

where $\Delta T = T - T_{\text{CMB}}$, and y_{col} ($y_{\text{Ly}\alpha}$) encode the ratio of collision (UV photon scattering) to absorption rate.

Equation (2) nicely demonstrates the evolution of the spin temperature relevant for the 21-cm physics. At early times, down to $z \sim 100$, collisions dominate, $y_{col} \gg 1$, $y_{Ly\alpha}$, and the spin temperature is that of the gas. At later times, CMB-induced absorptions followed by emissions begin to dominate, $y_{col,Ly\alpha} \to 0$, and the spin temperature rises above the gas temperature and toward that of the CMB. Finally, when the Ly-α-induced collisions are largest, $y_{Ly\alpha} \gg 1$, y_{col}, at around $z \simeq 20$, the spin temperature follows $T_{Ly\alpha}$. Since the Ly-α radiation can only be hotter than the gas and since the gas is colder than the CMB temperature, one concludes that the spin temperature cannot be lower than that of the gas.

A recent measurement by EDGES of the brightness temperature [7],

$$T_{21} \equiv \frac{1}{1+z} \left(T_s - T_{CMB} \right) \left(1 - e^{-\tau} \right), \tag{3}$$

where τ is the optical depth, found [5]

$$T_{21}^{EDGES}(z \simeq 17) = -500_{-500}^{+200} \, mK. \tag{4}$$

Here the errors correspond to the 99% C.L. intervals, and assuming the optimal scenario in which $T_s = T_{gas}$, the above implies $T_{gas}(z = 17) = 3.26_{-1.58}^{+1.94}$ K to be compared with the expected temperature, $T_{21}^{SM}(z = 17) \gtrsim -220 \, mK$. The discrepancy between expected and measured temperatures correspond to a 3.8σ excess.

An inspection of Eq. (3) reveals several possibilities that can address this discrepancy:

– Nonstandard CMB spectrum from astrophysical or new physics sources.
– Suppressed spin temperature from enhanced gas cooling or direct spin cooling.
– Suppressed optical depth.

Here we investigate the possibility that DM-gas interaction underlies the low gas temperature.

2.2 Dark Cooling

Since DM is significantly colder, it is naively expected to cool the gas down through its interaction [8]. However, the predicted relative bulk velocity between the two gases dissipates with time and thus, under certain conditions, act to heat up the gas [4, 9]. This competing effect is best seen through the Boltzmann equations describing the evolution of temperatures and of the relative velocity. Focusing on that of the gas temperature,

$$\frac{dT_{gas}}{d \log a} = -2T_{gas} + \frac{\Gamma_C}{H}(T_{CMB} - T_{gas}) + \frac{2}{3} \sum_{I=\{H,He,e,p\}} \frac{\dot{Q}_{gas}^I}{H}. \tag{5}$$

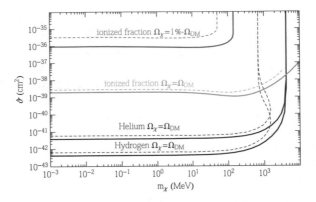

Fig. 1 The cross section required to fit the EDGES signal for DM-hydrogen interactions (**red**), DM-helium interactions (**blue**) and interactions with the ionized fraction assuming the interacting particle constitutes all of the DM (**green**) or only 1% of the DM density (**brown**). The solid (dashed) lines correspond to the *minimal* cross section needed to obtain a brightness temperature $T_{21} = -300$ mK $(-500$ mK) assuming infinite Ly-α radiation rate which couples the spin temperature to that of the gas, and assuming no heating of the gas due to X-ray radiation

The first term describes the usual redshift due to the expansion of the universe, the second term encodes the Compton scattering which acts to couple the gas to the photon bath, and the third describes the DM cooling term which schematically can be written as

$$\dot{Q}_{gas}^I \sim -x_i \Gamma_{\chi I} \Delta E_I + \frac{d}{dt} \frac{\mu_{\chi I} v_{rel}^2}{2}. \tag{6}$$

Here the first term describes cooling while the second heating due to relative velocity dissipation.

The above have two implications: (1) In order to beat the Compton scattering rates, the DM-gas scattering cross section must be large at $z \sim 20$ and (2) for the cooling term to dominate over the heating term, DM must have a sub-GeV mass. A large DM-baryon cross section is highly constrained and to allow for significant interactions at early times, Coulomb-like interactions must be assumed: $\sigma^I = \hat{\sigma}^I v_{rel}^{-4}$. Figure 1 shows the required DM-gas cross sections in order to address the EDGES anomaly.

2.3 Constraints

The large DM-gas cross sections shown in Fig. 1 require a light force carrier to communicate the interactions. Two possibilities exist: unscreened and screened forces.

Consider first models where a new light mediator induce Coulomb-like interactions between DM and hydrogen or helium (as apposed to their constituents). In this

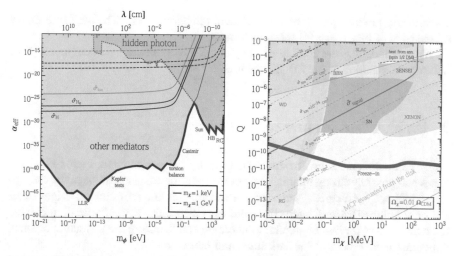

Fig. 2 **Left**: Constraints on the effective couplings of light mediators to the SM as a function of the mediator mass. The **gray shaded** region is excluded for theories that mediate a long-range Rutherford-like force which cannot be fully screened. For the astrophysical bounds, we assume democratic mediator couplings between electrons and protons. The **purple-shaded** region holds also for a light hidden photon under which SM charges are proportional to their electric charge and can therefore be screened. The **solid (dashed)** lines indicated the *minimal* α_{eff} needed to fit the EDGES signal (see Fig. 1) when cooling via the scattering of a keV (GeV) DM with hydrogen (**red**), helium (**blue**) and free electrons and protons (**green**) is assumed. **Right**: Constraints on the charge, Q, of a millicharged particle as a function of the DM mass. The **red** line indicates the minimal cross section needed to explain the EDGES measurement, assuming the millicharged particle constitutes only 1% of the DM density. See [4] for more details

type of scenarios, the mediator mediates a new unscreened long-range force. The strength of that force is described by the effective Yukawa potential

$$V(r) = \frac{\alpha_{eff}}{r} e^{-m_\phi r}. \tag{7}$$

On the left of Fig. 2 we show the limits (total shaded region) on such a mediator in the SM effective coupling verses mediator mass plane, alongside the parameters needed to explain the EDGES signal for keV and GeV DM mass. The red, blue, and green lines indicate the needed couplings assuming gas cooling via hydrogen, helium, and ionized fraction, respectively. We see that such a possibility is excluded as an explanation to the anomaly.

Models with screened forces include the hidden photon model in which the DM is charged under a $U(1)_D$ gauge group which kinetically mix with the SM photon, and a millicharged dark matter. In both cases, dark matter must cool the free electrons and protons directly as its interaction with hydrogen and helium is screened. The former case can be shown to be completely excluded [4] while the second has only a small region of allowed parameter space for a bosonic millicharged dark matter that constitutes no more than 1% of the dark matter in our universe. The limits are shown on the right of Fig. 2. See [4] for more details.

We conclude that the cooling of the gas via the dominant DM component is an unlikely explanation of the EDGES anomaly.

3 Probing Dark Matter with AGN

Many production mechanisms studied in recent years point to dark sector with light states and possibly strong interactions. Such scenarios are also motivated by the various small-scale structure discrepancies mentioned in the introduction. Light states not only induce DM self-interactions but may also introduce a dissipation mechanism that allows for loss of angular momentum and therefore the formation of small structures such as dark disks, dark stars, and more.

Various constraints imply that such dark structure is constrained to be less than 1% of the DM density. It is therefore interesting to ask whether such a dissipative component be discovered despite being very weakly coupled to the visible sector. In this section, we make progress in answering this question by considering dark matter accretion in Active Galactiv Nuclei (AGN). The details of this work are described in [6].

3.1 AGN Basics

AGN are currently understood as Super Massive Black Holes (SMBHs) at the centers of galaxies, which are undergoing an active phase of accretion of matter [10]. Baryonic matter forms an accretion disk that surrounds the BH and by losing angular momentum, this matter falls into the BH, feeding it, while at the same time releasing radiation to the surroundings. SMBHs are the result of accretion onto seed BHs, which are formed at high redshift. All known mechanisms for seed formation predict BH masses below (typically much below) $10^6\ M_\odot$ [11, 12].

A simple accretion picture consists of a thin disk which contains ionized gas which, due to viscosity, loses angular momentum and falls toward the BH. As matter falls into the BH, part of its gravitational potential energy is converted into radiation, which is observed as the AGN luminosity,

$$L = -\eta \dot{M}_{\mathrm{disk}} . \tag{8}$$

Here η is the radiative efficiency which ranges from 0.057 to 0.42 [13] and $\dot{M}_{\mathrm{disk}}$ is the time derivative of the disk mass. Under certain conditions, the luminosity is bounded from above by the Eddington luminosity,

$$L_{\text{Edd}} = 4\pi G_N \frac{M_{\text{BH}} m_p}{\sigma_T}, \tag{9}$$

where G_N is Newton's constant, M_{BH} the BH mass, m_p the proton mass and σ_T the Thomson scattering cross section. The time-averaged BH accretion rate can be written as

$$\langle \dot{M}_{\text{BH}} \rangle = -(1-\eta)\langle \dot{M}_{\text{disk}} \rangle = \frac{1-\eta}{\eta} \frac{L}{L_{\text{Edd}}} \frac{M_{\text{BH}}}{\tau_{\text{Sal}}} D, \tag{10}$$

where

$$\tau_{\text{Sal}} \equiv \frac{\sigma_T}{4\pi G_N m_p} \simeq 4.5 \times 10^8 \text{ yr} \tag{11}$$

is the Salpeter time [14] and $0 < D < 1$ is the so-called duty cycle. Assuming a constant L/L_{Edd} one then finds,

$$\log\left(\frac{M_{BH}}{M_{\text{seed}}}\right) = \frac{(1-\eta)}{\eta} D \frac{L}{L_{\text{Edd}}} \frac{t}{\tau_{\text{Sal}}}. \tag{12}$$

3.2 Observations

Several measurements of AGN exist. Here we use a sample of 40 AGN measured at redshift $z \sim 4.8$ [15]. In this sample, the mean BH mass is $\sim 8 \times 10^8 M_\odot$ and the mean value for L/L_{Edd} is ~ 0.6.

Using the above results, and setting $\eta = 0.1$, one can trace back the required seed BH mass, at $z \sim 20$, needed to explain the observation. The result, shown in Fig. 3, demonstrates the problem: assuming (sub-) Eddington accretion (as is typically the case) requires anomalously large seed masses in many of the observed AGN. Possible periods of super-Eddington accretion or additional growth through BH-BH mergers can ameliorate the tension present in the naive picture discussed above. Nevertheless, it is worthwhile to explore other avenues in order to address the puzzle. Below we study how dissipative dark matter could form a dark accretion disk and contribute to fueling the growth of SMBHs.

3.3 Dark Accretion

The presence of a dissipative dark component can strongly influence the evolution of BH growth rate. For concreteness, we assume below a simple scenario in which a dark electron, e', and a dark proton, p', interact via a hidden $U(1)$ gauge group.

Several conditions must be met to allow for dark accretion:

– Efficient accretion, allowing for a viscosity mechanism and fast growth rate, $\tau_{\text{Sal}}^h \ll \tau_{\text{univ}}$.

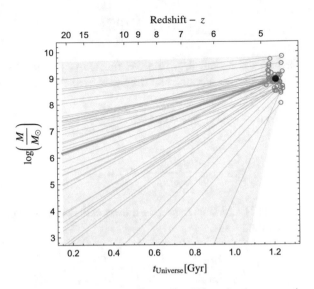

Fig. 3 Time evolution of BH masses according to Eq. (12), under the assumption of $\eta = 0.1$ and duty cycle $D = 0.5$, for the sample of 40 AGN analyzed in Ref. [15]. The *gray circles* represent the measured BH masses. The *black dot* is the sample's mean mass. The *thin green curves* show the evolution of each SMBH in the sample. Their slopes are dictated by the ratios L/L_{Edd} measured for each BH, which are assumed to be constant throughout the evolution. The *thick green curve* corresponds to the mean value of the sample, $L/L_{Edd} \approx 0.6$. Note that, in this simple picture, some BHs require seed masses much larger than $10^6 \, M_\odot$ at $z = 20$. The *yellow shaded region* represents the region enclosing all possible BH growth histories if the duty cycle is allowed to vary between 0.1 and 1

- Formation of a bound substructure via the cooling of the DM gas which subsequently falls into an accretion disk, $\tau_{cool} \ll \tau_{univ}$.
- Large duty cycle in which the formation of an accretion disk is much faster than the accretion time, $\tau_{cool} \lesssim \tau_{acc} \ll \tau_{univ}$.

Assuming, first, the presence of a dark accretion disk, it is straightforward to extend the standard analysis to include a dark sector. One finds,

$$\log \left(\frac{M_{BH}}{M_{seed}} \right) = \frac{(1 - \eta)}{\eta} \left(D^v \frac{L^v}{L^v_{Edd}} + D^h \zeta \frac{L^h}{L^h_{Edd}} \right) \frac{t}{\tau_{Sal}}, \tag{13}$$

where L^h_{Edd} is the hidden-sector Eddington luminosity, $\zeta \equiv \frac{L^h_{Edd}}{L^v_{Edd}} = \frac{\sigma_T/m_p}{\sigma'_T/m_{p'}}$, σ'_T is the hidden-sector equivalent of the Thomson cross section, $\sigma'_T = (8\pi/3)(\alpha'^2/m^2_{e'})$, and D^h is the hidden duty cycle. The shaded blue regions in Fig. 4 show the allowed parameter space in which dissipative DM accretion together with baryonic matter accretion can account for all the 40 BH masses starting from various maximal seed BH masses between 10^2–$10^6 \, M_\odot$ at $z = 20$ (different opacities correspond to different seed masses).

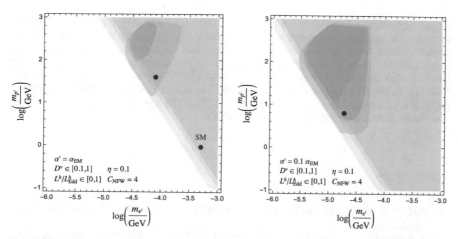

Fig. 4 Consistent region of parameters for a characteristic choice of $\eta = 0.1$ and $C_{\text{NFW}} = 4$. **Left panel**: The dark coupling is set to $\alpha' = \alpha_{\text{EM}}$. The *red dot* represents the SM values $(\alpha_{\text{EM}}, m_e, m_p)$. **Right panel**: The dark coupling is set to $\alpha' = 0.1\alpha_{\text{EM}}$. In the *shaded blue regions* the tension in the measured SMBH mass is resolved under the naive assumptions in regards to the presence of an accretion disk, while in the *shaded green regions* the conditions to form an accretion disk are also met. The *increasingly opaque blue and green regions* represent the regions in which we assumed different maximal seed BH mass: lightest for $M_{\text{seed}} = 10^6 \, M_\odot$, medium for $M_{\text{seed}} = 10^4 \, M_\odot$ and darkest for $M_{\text{seed}} = 10^2 \, M_\odot$

This simple analysis is based on the assumption that a dark accretion disk exists continuously, with enough DDM at its disposal to fuel the BH growth. The conditions to form and sustain a dark accretion disk are more involved and go beyond the scope of this note. We refer the reader to the original study [6]. Once taken into account, the available parameter space shrinks from the blue to the green shaded regions shown in Fig. 4 (again with the varying opacities corresponding to different maximal seed masses).

We conclude that dissipative DM can indeed influence the BH growth rate in an observable manner. Future dedicated studies of this kind and progress in the understanding of accretion disks may point to the presence of such DM via these indirect observations.

4 Outlook

It is possible that dark matter may not be discovered in conventional ways due to its weak subtle interactions with the visible sector. Nonetheless, as we discussed here, indirect effects of DM on structure formation at small and large scales may point to a nontrivial complex dark sector, allowing one to draw detailed conclusions in regards to the particle identity of dark matter. Much more work is needed in this direction.

Acknowledgements I would like to thank the organizers of this symposium and the Simon's foundation for producing this unique meeting. I would also like to thank my collaborators on these projects: Rennan Barkana, Nadav Outmezguine, Oren Slone, Diego Redigolo, Walter Tangarife, and Lorenzo Ubaldi. This work is supported in part by the I-CORE Program of the Planning Budgeting Committee and the Israel Science Foundation (grant No. 1937/12), by the Israel Science Foundation-NSFC (grant No. 2522/17), by the German-Israeli Foundation (grant No. I-1283-303.7/2014), by the Binational Science Foundation (grant No. 2016153) and by a grant from the Ambrose Monell Foundation, given by the Institute for Advanced Study.

References

1. D.N. Spergel, P.J. Steinhardt, Phys. Rev. Lett. **84**, 3760 (2000). https://doi.org/10.1103/PhysRevLett.84.3760
2. S. Tulin, H.B. Yu, K.M. Zurek, Phys. Rev. D **87**, 115007 (2013). https://doi.org/10.1103/PhysRevD.87.115007
3. M. Kaplinghat, S. Tulin, H.B. Yu, Phys. Rev. Lett. **116**(4), 041302 (2016). https://doi.org/10.1103/PhysRevLett.116.041302
4. R. Barkana, N.J. Outmezguine, D. Redigolo, T. Volansky, Phys. Rev. D **98**, 103005 (2018). https://doi.org/10.1103/PhysRevD.98.103005
5. J.D. Bowman, A.E.E. Rogers, R.A. Monsalve, T.J. Mozdzen, N. Mahesh, Nature **555**(7694), 67 (2018). https://doi.org/10.1038/nature25792. URL http://www.nature.com/doifinder/10.1038/nature25792
6. N.J. Outmazgine, O. Slone, W. Tangarife, L. Ubaldi, T. Volansky, JHEP **1811**, 005 (2018). https://doi.org/10.1007/JHEP11(2018)005
7. P. Madau, A. Meiksin, M.J. Rees, Astrophys. J. **475**, 429 (1997). https://doi.org/10.1086/303549
8. H. Tashiro, K. Kadota, J. Silk, Phys. Rev. D **90**(8), 083522 (2014). https://doi.org/10.1103/PhysRevD.90.083522
9. J.B. Muñoz, E.D. Kovetz, Y. Ali-Haïmoud, Phys. Rev. D **92**(8), 083528 (2015). https://doi.org/10.1103/PhysRevD.92.083528
10. H. Netzer, *The Physics and Evolution of Active Galactic Nuclei* (2013)
11. D.M. Alexander, R.C. Hickox, New Astron. Rev. **56**, 93 (2012). https://doi.org/10.1016/j.newar.2011.11.003
12. M.A. Latif, A. Ferrara, Publ. Astron. Soc. Austral. **33**, e051 (2016). https://doi.org/10.1017/pasa.2016.41
13. S.L. Shapiro, Astrophys. J. **620**, 59 (2005). https://doi.org/10.1086/427065
14. E.E. Salpeter, Astrophys. J. **140**, 796 (1964). https://doi.org/10.1086/147973
15. B. Trakhtenbrot, H. Netzer, P. Lira, O. Shemmer, Astrophys. J. **730**, 7 (2011). https://doi.org/10.1088/0004-637X/730/1/7

Halometry from Astrometry: New Gravitational Methods to Search for Dark Matter

Neal Weiner

Abstract Time domain astronomy offers the possibilty of news lensing searches. By looking for dramatic proper motions, measureable changes in them, or correlations between them, we can infer or constrain the presence of dark objects in our halo, such as black holes, subhalos, or other exotic objects. We consider new search strategies and the possibilities for current and future experiments.

Keywords Lensing · Dark matter · Black holes

1 Introduction

With the tremendous success of the standard model, it remains an open question what will take us to the next step in moving beyond it. So far, the LHC has not shown any sign of new physics. At the same time, it is clear that there must be physics beyond the standard model, evidenced by the existence of gravity, neutrino masses, dark matter, inflation in the early universe, and other signs. Of these, dark matter is perhaps the most promising in that there is reason to hope it may be connected to a physics scale low enough to be detectable by existing experiments. The ubiquitous presence of dark matter throughout the universe also gives great hope that we can test at least some of its properties directly.

Recently there have been a wide set of new ideas to look for dark matter [1]. As WIMP search experiments have reached maturity, a great deal of research has begun searching for new ideas to go to mass ranges well outside the weak scale. The range of new ideas has offered the prospect of testing alternative thermal models and in the MeV to GeV range as well as testing models even at much lower math skills such as 10^{-20} eV.

N. Weiner (✉)
Department of Physics, Center for Cosmology and Particle Physics,
New York University, New York, NY 10003, USA
e-mail: neal.weiner@nyu.edu

© Springer Nature Switzerland AG 2019
R. Essig et al. (eds.), *Illuminating Dark Matter*, Astrophysics and Space
Science Proceedings 56, https://doi.org/10.1007/978-3-030-31593-1_20

At the same time there have been a number of new ideas of have to pursue gravitational searches in the properties of dark matter (see work by Dalal, this volume). Gravitational probes of dark matter, while not probing directly its particle nature, cast a wide net and allow us to understand in broad brush the properties of the dark matter. Indeed, the successes that we have so far had in determining properties of dark matter have all come from gravitational searches. The fact the dark matter is cold, the fact the dark matter has scale invariant of density perturbations, the fact that dark matter is at least for the most part not self interacting, all these have come from gravitational searches for dark matter.

The purpose of this work is to consider new ideas to look for dark matter with lensing. It summarizes the results of [2], where further details can be found. Lensing of course is not new (see e.g., [3–5]). It is been used for decades to determine the presence and properties of dark matter especially around galaxy clusters. We shall be looking at dark matter in the time domain. This, too, is not new, as people of studied the possibility of dramatic lensing events when a dark object passes in front of a luminous source [6–11]. What is new is the high precision of the data and the enormous number of targets, which together open up new avenues to look for lensing it dark matter.

The methods in this work will be used to look for regions of high matter density in and around the Milky Way. Before discussing the precise techniques it is worthwhile to consider what sorts of things we might have a hope of finding. Almost certainly there are normal dark matter halos that are too small to host luminous sources. These halos at $10^8 M_\odot$ and below are expected to be common in the Milky Way Halo. In addition to this, we expect there to be a large number of black holes as well. These could arise from many sources, via conventional processes and mergers, from primordial processes or collapse of dark objects. These black holes could come in a wide range of masses and constitute an unknown fraction of the dark matter.

Density perturbations in the early universe while scale invariant over the scales we have perceived may actually show dramatic enhancement at small scales. As an example, we show in Fig. 1 limits on the *primordial* density perturbation from various sources. We overlay the expected size of perturbations if one takes the best fit value for n_s as well as its first and second derivatives from Planck [12] and extrapolate to smaller scales. As one sees, there can be a significant enhancement of $\delta\rho/\rho$ at smaller scales, even with simple evolution of the perturbations. Of course when shouldn't take the first and second relatives from Planck seriously at these scales, where the third, fourth and fifth derivatives would also become important. This merely shows that one needn't invoke some abrupt change in the properties of the perturbations in order to have a dramatic effect at small scales.

The dynamics of dark matter can also lead to high-density objects. Dissipative dynamics, dark stars and more all motivate the possibility of high-density regions, new compact objects and other significant lensing sources that could be present in large numbers throughout the Milky Way. And although we shall not discuss here, there is the exciting possibility of looking for dark solar system objects as well.

An important first question to ask is: what is changing? Most recently, GAIA [13] it has published its first release of data with 5D astrometric solutions. GAIA is an

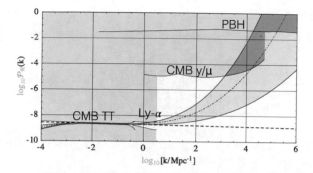

Fig. 1 Constraints on the primordial curvature power spectrum $\mathcal{P}_\mathcal{R}$ as a function of comoving wavenumber k (in units of Mpc^{-1}). Gray regions are excluded at 95% CL by temperature anisotropies in the cosmic microwave background (CMB TT), Lyman-α observations, nondetection of spectral distortions of $y-$ and μ-type in the CMB, and limits on primordial black holes (PBH). The black dashed line is the best fit to the *Planck* CMB data assuming a constant spectral tilt n_s, while the blue band indicates the parameter space where $dn_s/d \ln k$ and $d^2 n_s/(d \ln k)^2$ were allowed to float by 1σ from their best fit values (dot-dashed blue). Figure adapted from [2]

optical space based mission that will achieve precision $O(100\,\mu\text{as})$ over it several years of operation. It will see 1 billion stars at this level of precision, opening up the new possibility of understanding the properties of the Milky Way statistically. In the future we expect additional follow-up missions such as THEIA [14] to improve sensitivity by factor of roughly order 10. The Square Kilometer Array, a radio telescope [15], offers the prospect of measuring 10^7–10^8 quasars to accuracy in position of $O(10\,\mu\text{as})$.

2 A Brief Review of Weak Lensing

Strong lensing can occur when a luminous source comes within an Einstein radius θ_E of a gravitating body. Strong lensing yields multiple bright images and dramatic distortions of the source. If the source is farther than θ_E, then weak lensing occurs. For our purposes, the dominant effect of weak lensing is to shift the apparent location of the source.

The schematics are laid out in Fig. 2. The relevant quantities are the distance to the luminous source D_i, the distance to the lens D_l, the impact parameter from the path from the source to the observer b_{il} and the mass of the source M_l.

The size of the angular shift is $\Delta\theta = \frac{4G_N M_l}{b_{il}}$. This shift can be quite small—for a solar mass black hole and a near approach of $b \sim 10^{-3} pc$ we have a shift of $\sim 40\,\mu\text{as}$, below the detectable level for most stars. A $10^8 M_\odot$ halo with light passing by the edge of its scale radius $b \sim 1\text{kpc}$, yields a shift of $\sim 4\,\text{mas}$.

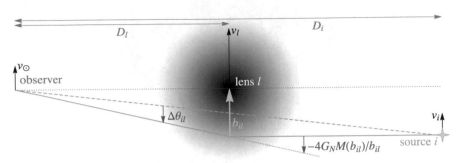

Fig. 2 A schematics of astrometric weak lensing geometry. See text for details. Figure taken from [2]

These shifts are in general not observable, because the position of an object is not known a priori. However, in the time domain, there are new possibilities. As the lens moves across the sky relative to the source, this position can change, leading to an apparent motion. This apparent motion yields the prospect for new searches.

3 Lensing in the Time Domain

There are essentially two relevant regimes of interest—situations where $\delta b/b < 1$ and $\delta b/b > 1$. When $\delta b/b > 1$ one has what we refer to as a "blip." The lensed image—progressing on what was otherwise a straight line, deviates by a significant amount and then returns to its straight line trajectory. Such dramatic events have been studied previously [6–11]. Here we will attempt to quantify what such events can yield in the search for dark matter.

In the regime where $\delta b/b$ is small, there is no dramatic event and this regime has not been previously explored. Here, one can expand the apparent motions in powers of $\delta b/b \sim v_{il}/b$, where v_{il} is the motion of the lens relative to the source, so $\Delta\dot{\theta} \sim v\mathsf{b} \times \Delta\theta$ and $\Delta\ddot{\theta} \sim (v\mathsf{b})^2 \times \Delta\theta$. These apparent velocities and accelerations are often quite small, but critically they have well defined patterns. For an NFW halo, one has a somewhat different pattern, but which, for impact parameters $b > r_s$, returns to the pattern for a point source mass.

The size of these effects can often be extremely small. However, with the large number of objects under discussion, there are a number of new possibilities.

We can categorize these roughly into two groups: first are *Rare objects*—in the presence of many objects, close encounters become possible. Examples of this would be

- **Blips**—A lens passes near a source with $\delta b/b \sim 1$, leading to a pronounced deflection.
- **Outlier**—$\delta b/b < 1$ but lens is sufficiently close that acceleration or velocity is above expectations. This could be, for instance, hypervelocity stars, or stars with dramatic accelerations.

To accurately determine a blip, we would need a more complete time series of positions than currently provided by GAIA. At the moment, only 5D astrometric solutions are provided (parallax, angular position and proper motion). Velocity outliers can be studied with the current data, but they are perhaps the most problematic source, since nearby stars with poorly measured parallax can contaminate the sample and produce tails that will limit sensitivity. That said, lensing provides an interesting possible explanation for bizarre hyper-velocity stars, such as SDSS J121150.27+143716.2 [16], a binary system that would have been disrupted by ordinary acceleration mechanisms. Acceleration, an observable with less intrinsic noise, is more promising, but awaits future data releases with more general astrometric solutions.

The second category directly exploits the signal-to-noise statistics from multiple luminous sources. These are *multi-object observables*—situations where effects on individual sources are below noise, but in aggregate can be visible. Examples would be

- **Multi-blips**—A lens passes near many sources with $\delta b/b \sim 1$.
- **Templates**—small effects are matched to an expected pattern arising from a diffuse halo.
- **Correlations**—no large scale patterns, but "nearby" sources have motions or accelerations correlated with each other.

As with blips, multi-blips require more complete time-series information, but offer the prospect of detecting massive solar system objects as well as halo ones [2].

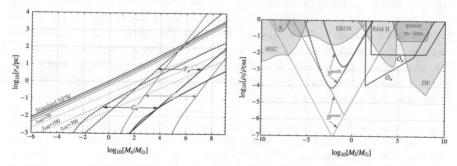

Fig. 3 Sensitivity projections (S/N = 1) for techniques described in this paper. Solid curves show expected sensitivity with near term data, while dashed curves show projections for future experiments. *Left*: Sensitivity projections for NFW subhalos as a function of core mass M_s versus scale radius r_s. The blue curves show sensitivities for the template velocity test statistic. The green curves show the global velocity correlation test statistic sensitivity. The red curves depict the sensitivity for the global acceleration correlations. Also shown in solid gray is the "standard" NFW subhalo median relation between M_s and r_s from Ref. [17] for three subhalo distances $R_{\rm sub} = \{240, 10, 5\}$ kpc away from the Galactic Center (closer ones are denser), as well as estimates (dotted gray) for nonstandard collapse redshifts $z_{\rm coll}$. *Right*: Projected sensitivity of point-like lensing searches for compact objects of mass M_l and dark-matter fraction ρ_l/ρ_{DM}. In thin and thick solid orange, we show unit signal to noise lines using mono- and multi-blip searches; blue dashed and red dotted lines depict fiducial sensitivity from outlier velocity and acceleration events. Also shown by the gray regions, from left to right, are constraints from Kepler, Subaru HSC, and EROS microlensing, the survival of the Eridanus II cluster, quasar millilensing, and dynamical friction. See [2] for details, from where these figures are taken

Correlations of velocities require lower noise than found with stars, and we would use instead quasars. A future mission such as SKA could provide accurate measurements of these to achieve good reach. For stars, acceleration correlations provide sensitivity, but likely will await future astrometric missions to yield strong reach.

In the immediate term, templates provide the best prospect of limits. These are matched filters to the precise dipole-like pattern expected to be produced by a moving lens. This relies only on having accurate velocity measurements and aggregating the data from many stars.

These different techniques offer different sensitivity at different masses or to point vs diffuse sources. We sketch out the expected sensitivity in Fig. 3. We show the expected sensitivity from GAIA, but also from future missions, such as THEIA or SKA.

4 Discussion

Although lensing is now a mature area, time domain lensing is in its infancy. However, because of the dramatic improvement in data that we have seen and can expect, there is a real prospect to constrain or detect a wide range of dark objects in the Milky Way halo.

What is perhaps most interesting about time-domain lensing is that we can use time as an incredible lever arm to gain sensitivity. Although it may extend outside our lifetimes, precision measurements of e.g., quasars at the $10\,\mu$as level done over half a century could yield incredible knowledge about their apparent motion, which, in turn, could provide great insight into the properties of dark matter. It is worth remembering that it took us fifty years to discover the Higgs boson, so it could well take five hundred to get a similar level of understanding about the dark sector. It is worth thinking now about studies with long time horizons that can make progress regardless of what the future brings.

Acknowledgements This research was supported in part by the National Science Foundation under Grant No. NSF PHY-1748958 and PHY-1620727. The work of NW is supported by the Simons Foundation. This work has made use of data from the European Space Agency (ESA) mission *Gaia* (https://www.cosmos.esa.int/gaia), processed by the *Gaia* Data Processing and Analysis Consortium (DPAC, https://www.cosmos.esa.int/web/gaia/dpac/consortium). Funding for the DPAC has been provided by national institutions, in particular the institutions participating in the *Gaia* Multilateral Agreement.

References

1. M. Battaglieri, et al., (2017)
2. K. Van Tilburg, A.M. Taki, N. Weiner, JCAP **1807**(07), 041 (2018). https://doi.org/10.1088/1475-7516/2018/07/041
3. Y. Mellier, Ann. Rev. Astron. Astrophys. **37**, 127 (1999). https://doi.org/10.1146/annurev.astro.37.1.127

4. N. Dalal, C. Kochanek, Astrophys. J. **572**(1), 25 (2002)
5. C. Kochanek, N. Dalal, Astrophys. J. **610**(1), 69 (2004)
6. M. Hosokawa, K. Ohnishi, T. Fukushima, M. Takeuti, Astron. Astrophys. **278**, L27 (1993)
7. E. Hog, I.D. Novikov, A. Polnarev, Astron. Astrophys. **294**, 287 (1995)
8. M. Miyamoto, Y. Yoshii, Astron. J. **110**, 1427 (1995)
9. M.A. Walker, Astrophys. J. **453**, 37 (1995)
10. A. Boden, M. Shao, D. Van Buren, Astrophys. J. **502**(2), 538 (1998)
11. M. Dominik, K.C. Sahu, Astrophys. J. **534**(1), 213 (2000)
12. P. Ade, N. Aghanim, M. Arnaud, F. Arroja, M. Ashdown, J. Aumont, C. Baccigalupi, M. Ballardini, A. Banday, R. Barreiro et al., Astron. Astrophys. **594**, A20 (2016)
13. T. Prusti, J. De Bruijne, A.G. Brown, A. Vallenari, C. Babusiaux, C. Bailer-Jones, U. Bastian, M. Biermann, D. Evans, L. Eyer et al., Astron. Astrophys. **595**, A1 (2016)
14. C. Boehm, A. Krone-Martins, A. Amorim, G. Anglada-Escude, A. Brandeker, F. Courbin, T. Ensslin, A. Falcao, K. Freese, B. Holl, et al., (2017), arXiv preprint arXiv:1707.01348
15. M.J. Jarvis, D. Bacon, C. Blake, M.L. Brown, S.N. Lindsay, A. Raccanelli, M. Santos, D. Schwarz, (2015), arXiv preprint arXiv:1501.03825
16. P. Németh, E. Ziegerer, A. Irrgang, S. Geier, F. Fürst, T. Kupfer, U. Heber, ApJ **821**, L13 (2016). https://doi.org/10.3847/2041-8205/821/1/L13
17. Á. Moliné, M.A. Sánchez-Conde, S. Palomares-Ruiz, F. Prada, Mon. Notices R. Astron. Soc. **466**(4), 4974 (2017)

Complex Dark Sectors and Large Bound States of Dark Matter

Kathryn M. Zurek

Abstract We discuss models of hidden sector dark matter, and a recent generic proposal for large bound states of asymmetric dark matter.

1 Dark Matter Model Building in a Data-Driven Era

When is it worthwhile for a theorist to invest resources in conceiving a model of dark matter, deriving its consequences, and collecting results in the form of a clearly written paper? After all, the probability of any model being a correct description of our Universe is, at best, $\mathcal{O}(\epsilon)$. One can demand that the dark matter candidate solve some other problem. For example, the community has been largely focused over much of the last 35 years on two candidates, the WIMP and axion. These are viewed as being the theoretically most elegant, largely because they solve two problems instead of one: the dark matter problem and either the hierarchy problem or the strong CP problem, respectively. These two candidates have an advantage that they make well-defined and highly testable predictions, which the author believes should be pursued to the end.

It may not, however, be the job of the dark matter to solve a problem besides the dark matter problem. That is, the WIMP or the axion may not be the solution (or the *whole* solution) the Universe has chosen. In this case which guiding principle does one use? Already there is a zoo of candidates available for perusal and consideration—why add more? I will propose the following two simple criteria as a guide.

1. The model should give *qualitatively* new cosmological, astrophysical, or terrestrial signatures. This does not mean that in every case a new experiment must be designed to search for the candidate in question. In many cases data analysis techniques and search methods will, however, need to be repurposed to give an adequate description or constraint on the model.

K. M. Zurek (✉)
Lawrence Berkeley National Laboratory, Berkeley, CA, USA
e-mail: kzurek@berkeley.edu

© Springer Nature Switzerland AG 2019
R. Essig et al. (eds.), *Illuminating Dark Matter*, Astrophysics and Space
Science Proceedings 56, https://doi.org/10.1007/978-3-030-31593-1_21

2. *No Rube Goldberg Machines.* The new signatures should be obtained with no energy expended solely to circumvent the requirements of a consistent theory and existing experimental constraints.

In other words, we are looking for new *paradigms* with *qualitatively distinct signatures*.

I'll first discuss one of these paradigm shifts that has happened over the last (approximately) decade, the move from dark matter as a single, stable, weakly interacting mass particle to the idea that the dark matter is part of a hidden world with complex dynamics. I will then examine a simple(!) model of large bound states of dark matter that meets the above two criteria, generating qualitatively new features. This demonstrates that we have not yet entered the Rube Goldberg Machine era of dark matter model building.[1]

2 Complex Hidden Sectors

Dark sectors have become a set of buzz words in recent years. It is meant to signify that one is relaxing the requirement that the dark matter solve a second problem (like strong CP or hierarchy problems), and instead focus on the range of dynamics that the dark matter can exhibit consistent with our cosmological history and present knowledge of the dark matter. One keeps a keen eye on how to "be ready for anything" in terms of rooting out the nature of the dark matter. Even in the case of light states the cosmological and accelerator constraints are typically not very strong.

The essential point of the hidden valley paradigm is that dark and visible matter may come from two separate sectors, separate from any other problem one seeks to solve in the Standard Model. Light states in the dark sector are consistent with all cosmological, astrophysical and terrestrial constraints if they couple either through (a) states with mass in excess of the weak scale or (b) through light connector states with relatively small couplings (typically smaller than $\sim 10^{-2}$) to SM states. Further the dynamics of the hidden sector / hidden valley can be complex, complicating and enriching the nature of the signatures.

Take for example the case of a strongly coupled hidden sector with two light flavors, in the absence of both electromagnetism and the weak interactions [1]. The two dark quarks with bind into dark pions, π_v^+, π_v^-, π_v^0, where the superscripts denote the isospin charge. Because of the absence of dark weak interactions, isospin is conserved, and the π_v^+, π_v^- states are viable DM candidates, with π_v^0 decaying to pairs to SM states. If there is further a particle-anti-particle asymmetry between π_v^+ and π_v^-, the annihilation process

$$\pi_v^+ \pi_v^- \rightarrow \pi_v^0 \pi_v^0 \tag{1}$$

---

[1]One could argue that this is contrast to weak scale model building.

Freezes out when the antiparticle is depleted (and provided the motivation for Asymmetric Dark Matter in [2]). This model allows for a simple, generic cosmology that, because the states are light, weakly coupled and hence long-lived, generates novel terrestrial signatures. Terrestrial signatures of light long-lived particles from hidden valleys are now well studied (see for example the White Paper [3]). Such a model satisfies our criteria: it is quite simple and generic (decidedly not a Rube Goldberg Machine)—many types of models, strongly and weakly coupled; few flavor and many flavors; different gauge group structure, will give related types of novel signatures.

It immediately becomes clear, however, that the number of variations possible in models of dark matter approaches infinity. Given the massive development, and the growing numbers of papers, in this area, one must consider whether writing another variation on a hidden sector model is a worthwhile investment of time. We next consider a simple and generic model that we argue satisfies the two criteria above.

3 Big Composite Dark Matter States

Let's first consider bound states in the SM, known as nuclei. Here we follow a recent series of papers [4–6], along the same directions as [7, 8]. Synthesis of SM nuclei can be simply described through a freeze out equation

$$\frac{dN}{dt} = Nn_N\sigma_N v_N = N\sigma_0 N^{2/3} e^{-\alpha N^2/v_N} \frac{n_X}{N} v_N. \tag{2}$$

Here N represents a composite state traveling at velocity v_N having N nucleons, with a geometric cross section that grows according to the area of the composite, $\sigma_N = N\sigma_0 N^{2/3}$. We have also included a term to model the effect of a (putative) Coulomb barrier, $e^{-\alpha N^2/v_N}$. Let's take the synthesis at 0.1 MeV (due to the deuterium bottleneck) and calculate the size of the resulting nuclei:

$$N \approx \begin{cases} 2.6 & \text{With Coulomb barrier} \\ 10^4 & \text{Without Coulomb barrier.} \end{cases}$$

One immediately sees that in the absence of electromagnetism for dark matter, one synthesizes radically different size states.

Is this sensitive to the nature of the synthesis process? Consider a slightly different set-up: we presume the presence of a bottleneck, where an occasional bound state slips through the bottleneck with probability p. In this case the analogue of Eq. 2 is

$$\frac{dN}{dt} = kn_k\sigma_{kN} v_k, \tag{3}$$

where k represents the (vast majority) of small nuggets k that do not slip through the bottleneck, whereas N represents that rare capture site. In this case one finds

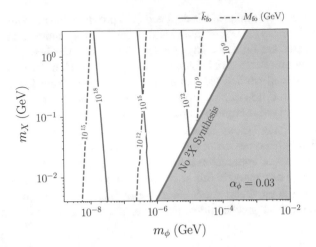

Fig. 1 Contours of typical nugget number exiting big bang darkleosynthesis, $\bar{k}_{\text{fo}}$ (dashed red) and typical nugget mass $\bar{M}_{\text{fo}}$ (solid purple) for $\alpha_\phi = 0.03$. The temperature of the dark sector is assumed to be roughly the same as the standard model photon temperature. The blue-shaded region corresponds to when the binding energy of the two-body state is smaller than the force mediator mass m_ϕ, where synthesis of the two-body state will not be efficient. The upper m_X cutoff corresponds to the requirement that two-body fusion rate is smaller than Hubble, and the lower m_X cutoff corresponds to requiring that synthesis happen before matter-radiation equality. The various kinks in the contours are results of the change in g_* as the synthesis temperature passes through QCD phase transition and neutrino decoupling. Figure from Ref. [5]

$$N \approx \begin{cases} 9 & \text{With Coulomb barrier} \\ 10^9 & \text{Without Coulomb barrier.} \end{cases}$$

Again one finds that one synthesizes much larger states with a simple modification that the nucleons do not couple to a massless, repulsive vector force.

So let's consider a simple hidden sector with a fermionic nucleon X, attractive scalar force ϕ and (massive) repulsive vector force V_μ. We will allow the mass of the nucleon, m_X, as the well as the masses of the forces m_ϕ and m_V vary. One can quite generally solve the equation of motion for this system and determine the size of the synthesized states under general conditions. The results are shown in Fig. 1, for the case of an attractive force only. As promised, one obtains *yuge!* bound states of dark matter. So clearly there is no Rube Goldberg machine to obtain these states. We call them, following [7], "nuggets."

Does this model meet our other criterion, that it gives rise to qualitatively distinct astrophysical, cosmological and terrestrial observables? Consider first the astrophysics and cosmology. When two nuggets fuse they behave just like clay putty—they form a composite state which thermalizes before settling down to its ground state via radiating force mediators or small nugget fragments (this is known as the "compound nucleus" model and well-reproduces the behavior of SM nuclei). This model implies that the interactions are *highly* dissipative, such that constraints from ordinary elastic self-interacting dark matter may not apply directly. One expects core contraction much more rapidly than in elastic SIDM, feeding the black hole at the center of the galaxy and leading to other modifications in astrophysics and cosmology. In addition, the large bound states of dark matter feature coherence that suggests that they may be searched for in experiments sensitive to low momentum transfer. With even a simple but non-trivial structure in the hidden sector, the cosmological implications and signatures are novel and open to further exploration.

References

1. M.J. Strassler, K.M. Zurek, Phys. Lett. B **651**, 374 (2007). https://doi.org/10.1016/j.physletb.2007.06.055
2. D.E. Kaplan, M.A. Luty, K.M. Zurek, Phys. Rev. D **79**, 115016 (2009). https://doi.org/10.1103/PhysRevD.79.115016
3. D. Curtin, et al. (2018). arXiv:1806.07396
4. M.I. Gresham, H.K. Lou, K.M. Zurek, Phys. Rev. D **97**(3), 036003 (2018). https://doi.org/10.1103/PhysRevD.97.036003
5. M.I. Gresham, H.K. Lou, K.M. Zurek, Phys. Rev. D **96**(9), 096012 (2017). https://doi.org/10.1103/PhysRevD.96.096012
6. M.I. Gresham, H.K. Lou, K.M. Zurek, Phys. Rev. D **98**, 096001 (2018)
7. M.B. Wise, Y. Zhang, Phys. Rev. D **90**(5), 055030 (2014). https://doi.org/10.1103/PhysRevD.90.055030, https://doi.org/10.1103/PhysRevD.91.039907. [Erratum: Phys. Rev. D91, no.3,039907(2015)]
8. E. Hardy, R. Lasenby, J. March-Russell, S.M. West, JHEP **06**, 011 (2015). https://doi.org/10.1007/JHEP06(2015)011